Achieving Excellence in the Fire Service

Judy Janing PhD., R.N., EMT-P

Gordon Sachs MPA, EFO

Upper Saddle River, New Jersey 07458

Library of Congress Cataloging-in-Publication Data

Janing, Judy.
Achieving excellence in the fire service / Judy Janing, Gordon Sachs.
p. cm.
Includes bibliographical references and index.
ISBN 0-13-042208-8
1. Fire Departments—Management. 2. Rescue work—Quality control. 3. Fire extinction—Quality control. 4. Customer services. I. Sachs, Gordon M. II. Title.

TH9158.J36 2003
363.37'068—dc21 2002032380

Publisher: *Julie Levin Alexander*
Publisher's Assistant: *Regina Bruno*
Senior Acquisitions Editor: *Katrin Beacom*
Editorial Assistant: *Kierra Kashickey*
Senior Marketing Manager: *Tiffany Price*
Production Information Manager: *Rachele Strober*
Director of Production and Manufacturing: *Bruce Johnson*
Managing Production Editor: *Patrick Walsh*
Manufacturing Buyer: *Pat Brown*
Production Liaison: *Julie Li*
Production Editor: *Lynn Steines, Carlisle Publishers Services*
Creative Director: *Cheryl Asherman*
Cover Design Coordinator: *Christopher Weigand*
Cover Designer: *Christopher Weigand*
Compositor: *Carlisle Publishers Services*
Printer/Binder: *Von Hoffman Press*
Cover Printer: *Coral Graphics*

Pearson Education LTD.
Pearson Education Australia PTY, Limited
Pearson Education Singapore, Pte. Ltd.
Pearson Education North Asia Ltd.
Pearson Education Canada, Ltd.
Pearson Educación de Mexico, S.A. de C.V.
Pearson Education—Japan
Pearson Education Malaysia Pte. Ltd.
Pearson Education, Upper Saddle River, New Jersey

10 9 8 7 6 5 4 3 2 1
ISBN 0-13-042208-8

This book is dedicated to the memory of my mother, who not only taught me, but lived by, the philosophy of "being the best you can be." She inspired me throughout her life and her memory has inspired me for all the years I have been without her. Although she never held an "important" position, she was a true leader, touching many lives with kindness, recognition, encouragement, and love.

—J. J.

This book is dedicated to my parents, Fred and Cynthia. Were it not for them, I would not know what excellence is, nor would I have the inclination to strive to reach it.

—G. S.

Contents

Acknowledgments

I could not have written this book without the support and encouragement of my mentor, friend, and co-author Gordon Sachs, my husband Corky, and my sons Jeff and Jamie. As a retired firefighter/paramedic, Corky understood this project and the challenges faced by the fire service of today. My sons are my greatest accomplishments in life and have chosen two true heroic paths. Jeff's career is Navy and Jamie is a paramedic. Their idealism and love of country and profession have lifted my spirits during the down days and continue to motivate me to pursue my life's goals.

—J. J.

This work, and a lot of the other work I do, would not be possible if it were not for the support and assistance I receive from my friend, co-instructor, and co-author, Judy Janing. I would also like to acknowledge the support of my wife and children—Lisa, Adam and Brandy—who have always been there through good times and bad. Special recognition goes to members of Fairfield (PA) Fire & EMS who have worked with me in moving the department toward excellence. In particular, thanks to Deputy Chief Bill Morton, who spent countless hours working on various projects and programs and, in the process, established many benchmarks for other departments.

—G. S.

Reviewer List

David S. Becker
Castle Rock Fire and Rescue Department
Castle Rock, CO

W. David Bunce
Salt River Fire Department
Scottsdale, AZ

Steven Edwards
Maryland Fire and Rescue Institute
University of Maryland
College Park, MD

Ken Riddle
Las Vegas Fire and Rescue
Las Vegas, NV

About the Authors

Judy Janing, Ph.D., R.N., EMT-P is an experienced researcher and analyst, and serves as an Emergency Program Specialist with IOCAD Emergency Services Group. Ms. Janing has over 25 years experience in emergency response and over 10 years experience in education. She has served as the lead educational developer for the development or revision of several courses at the National Fire Academy and various universities. Ms. Janing was involved with the Omaha, Nebraska fire department's EMS program for many years. Her Ph.D. is in Community and Human Resources, specializing in evaluation methodology; her Masters Degree is in Adult Education. She has published several clinical, educational, and research articles in various peer journals, and has co-authored several texts. She is an adjunct faculty member at the National Fire Academy.

Gordon M. Sachs, MPA, EFO is Director of IOCAD Emergency Services Group and has served as Chief of Fairfield (PA) Fire and EMS. He is a former program manager with the Federal Emergency Management Agency's U.S. Fire Administration. He has over 25 years of emergency response experience, particularly in the area of incident command, training, and safety. He is an adjunct faculty member in the *EMS Management/Leadership, Command and Control and Emergency Response to Terrorism* curricula at the National Fire Academy. He has written many journal articles on EMS management/leadership and responder safety/health, and has authored or co-authored several books in these areas. He has a Masters Degree in Public Administration, a Bachelor of Science in Education, three Associate in Applied Science Degrees in Fire Science disciplines, and an Executive Fire Officer Certificate from the National Fire Academy.

Achieving Excellence in the Fire Service

Introduction

1 CHAPTER

◆ HISTORY OF SYSTEM EVALUATION IN THE FIRE SERVICE

Webster defines *quality* as "degree of excellence; superiority in kind; a distinguishing attribute." An often-used, more simplistic definition of *quality* is "conformance to standards." To different people, *quality* means many different things. It is based on culture, training and education, experience, and knowledge outside their own environment. For example, a "quality car" to one person may be one that runs well and gets good fuel mileage, while to another person it may be one with a luxury package that provides extra comfort. Quality, in other words, is a function of expectations and perceptions. What is perceived is what the customer receives, or "perception is reality."

In real life, quality is almost always defined from the customer's point of view. For example, someone who calls 9-1-1 because her daughter is having trouble breathing wants emergency personnel to get there quickly, know what they are doing, be courteous while they are there, and take care of the problem efficiently. These factors will define the "quality" of care from the customer's point of view. The caller likely will not care if help arrives in an ambulance, a fire truck, a police car, or an SUV, as long as it arrives quickly. If the team takes 20 minutes to arrive, or if they are rude while on the scene, the quality of service perceived by the customer will be poor. The customer likely will not know the difference between a first responder, an EMT, or a paramedic. As long as this person helps her daughter to breathe better, the perceived quality of care will be high.

W. Edwards Deming, a renowned authority on quality management and leadership, has written that quality management—managing the work environment to maximize both the external and internal expectations and perceptions of quality—is key to a quality organization. This approach is much more effective than "quality control," which looks at individual processes typically performed by the members. Deming states that members typically work within a system that is beyond their control. The system, not the individual worker's skills, determines how that worker performs. Only management can change the system because management is responsible for it. Often, members are punished for accurately reporting system results, and are rewarded for providing the answers that managers desire; therefore, part of the internal politics on the job is doing what makes one look good to the boss. This is often more important than striving to look good to the customers.

In the fire service, it is often difficult for a firefighter to suggest to a chief officer that there might be a better way to do things. For one thing, how can a "rookie" know more than an experienced "smoke-eater"? Then, typically, a chain of command must be followed, and the suggestion can be stifled anywhere in that chain (especially if someone is reluctant to "step on toes"). Further, even though there are national standards that provide the information needed to conduct certain operations or programs, some chief officers may be reluctant to implement these because they contradict "the way things have always been done." These are the types of "systems" that often stand in the way of quality, or excellence, in the fire service.

Another example might be fire inspections by engine companies. The department may require that a certain number of these inspections be performed each year, and that the company officer is responsible for getting them done. The department may not provide the training necessary to perform the inspections effectively, or the company may be too busy with responses or training to perform all of the inspections or to perform the inspections thoroughly. So, while the paperwork at the end of the year may indicate that the inspections have been completed, they may actually not have been, or, if they were completed, they may have been second-rate. Records might also show inspections have been completed just to protect the chief officer's reputation and to prevent disciplinary action. This is another example of a "system" that may stand in the way of quality.

Some of the fire departments recognized today as quality organizations have eliminated, or at least minimized, the roadblocks that stand in the way of improvements and advances. Other departments live by the axiom of "two hundred years of tradition unimpeded by progress." While tradition has an important place in the fire service, tradition alone cannot effectively or efficiently carry any department into the future. Again, perception is reality; if internal customers only see what is done within their own organization, they have no way to judge whether they are performing quality services. They have nothing to compare themselves to. In effect, they are being stymied by ignorance (not knowing any different) and apathy (not caring that there may be other ways of doing things).

Fortunately (or unfortunately, depending upon one's point of view), the public is neither ignorant nor apathetic about today's fire service. While the public may look at firefighters as today's heroes, television reality shows and on-the-scene, real-time news have provided the public with perceptions and expectations that fire departments should be striving to live up to. Similarly, the American fire service in general knows and understands the consensus standards-making process, and recognizes when other fire departments are not following national standards. For example, letters to the editor of major fire service trade journals regularly address the fact that photos in those journals depict firefighters "doing things wrong" or not in conformance with national standards. The public often recognizes much of this as well.

The problem with this is simple: when the public—the primary customers—perceive that their tax dollars or donations are being spent for a less-than-quality service, they become dissatisfied. That leads to many problems. For example, word of a department's poor quality—whether an accurate perception or not—can spread quickly. Data collected in a 1978 study by the White House Consumer Affairs Panel showed the following.

1. A satisfied customer tells three people.
2. A dissatisfied customer tells 11 people.
3. In one study, 13% of dissatisfied customers complained to over 20 people.
4. Consider that each of the 11 people who heard a "bad experience story" told 11 others.
5. Ninety-six percent of all unhappy customers never tell the company.
6. Of those who do complain, between 54 % and 70 % who feel that their complaint was resolved will do business again.
7. This increases to 95 % if the customers feel that the complaint was resolved quickly.

Such complaints cost money, even in this day of "do more with less." Research on quality management shows that poor quality costs the service: often 20% of service goes to fixing bad quality. Fire departments that have to spend time and money on damage control often do not have the time or money to move forward toward progressiveness, yet those that focus on quality often have extra funds to put toward new ideas or projects. Further, city and county managers are often reluctant to provide extra funding to departments or agencies that show a lack of concern for quality and improvement, and that often cost the city or county in terms of liability or settlements. The same can be said of elected officials or candidates for office, especially if support for an agency or department with a poor reputation might cost them votes.

Complaints from customers are not always based on poor system quality and do not always need to be addressed with wide-scale changes. Still, feedback from both internal and external customers can help a department move in the right direction. Stu Leonard, another quality management guru, put it best when he said, "Customers who complain are our best friends because they give us the opportunity to improve."

◆ CURRENT PRACTICES OF SYSTEM EVALUATION

Customer-focused organizations, including fire departments, create predictably positive experiences by continually striving to exceed their customers' expectations. Fire departments today do much more than extinguish fires once they start, and the public knows that. They expect more than that. Most of today's workers—the "Generation X" and "Generation ME" folks—want more than that. It is up to management to create and support an environment where this internal desire to exceed and external desire for quality services can be realized. This requires a balance of two factors: human resources and physical resources. Both must work together if success is to be achieved.

Five keys ensure that human and physical resources are balanced, resulting in distinctive quality service.

1. Listening, understanding, and responding to the evolving needs of customers.
2. Establishing a clear vision of what good service is, communicating that vision to everyone, and ensuring that service quality is important.
3. Establishing concrete standards of service and regularly measuring against those standards.

4. Hiring good people, training them in service delivery, empowering them to work on behalf of the customer, and ensuring that they have the tools and skills to meet the service standards.
5. Recognizing and rewarding high quality service; celebrating and acknowledging those who "go one step beyond" for customers.

Fire departments that recognize these keys are often the ones recognized by their peer agencies across the country as being progressive or models for other departments to follow.

Another important approach that coincides with the concept of modeling other departments is "benchmarking." This entails identifying the highest achievable standard, and then attempting to meet or exceed that standard. To identify this standard, fire department managers must go to all levels of the organization for input, and must get input from customers, from those outside the fire service (this is called *functional benchmarking*), from other fire departments, and from stakeholders. The fire service manager must "become" the customer.

National consensus standards, such as those produced by the National Fire Protection Association (NFPA), are often a good place to start the benchmarking process. NFPA standards typically are the "minimum standard"; that is, they cite the criteria or components that must be met or exceeded in specific areas. Three of the most influential, if not controversial, NFPA standards related to the fire service have been NFPA 1500, Standard on Fire Department Occupational Safety and Health Program; NFPA 1710, Standard for the Organization and Deployment of Fire Suppression Operations, Emergency Medical Operations, and Special Operations to the Public by Career Fire Departments; and NFPA 1720, Standard for the Organization and Deployment of Fire Suppression Operations, Emergency Medical Operations and Special Operations to the Public by Volunteer Fire Departments. These standards identify the minimum levels at which a fire department should operate in terms of safety and deployment when such levels are not established by other legal entities. There are compliance matrixes available for these standards in the form of checklists, which can then be turned into implementation plans. Departments can use these standards to identify what needs to be done in order to meet minimum standards, as well as those areas where the minimum is exceeded.

While national consensus standards often cite the minimum acceptable standard, the desire of local customers (internal and external) may be higher than the minimum. Or, they might not even know that such a minimum standard exists! Similarly, external customers may have moved into an area from a place with a more progressive fire department, and may expect that all fire departments are the same. Internal customers—fire officers and firefighters—may never have seen a fire department more than 50 miles from their home. Thus, education is an important part of this process as well. Some of the questions in this process might be: What are the minimum standards? What are similar size fire departments across the country doing? What are other industries doing? Can we adapt what they are doing to meet our needs?

Many fire departments have identified benchmarks—where they want to be, so to speak—and have used those as a starting point. They often practice the Kaizen concept—a theory of continuous improvement where a department practices slow, gradual accomplishments moving toward being the best they can be. They are, in effect, in a race for which there is no finish line. While this can be very effective, there is one caution: do not equate the theory of Continuous Quality Improvement with the concept of "It's not good enough," especially if using committees. Instead, consider using the concept, "Can it be made better? If so, how?"

One fire department in Florida made the "never-good-enough" mistake—establishing CQI committees for many different areas—and once a plan for that area was developed and implemented, a new committee was established to improve upon that plan. Because all members knew that whatever their committee came up with would not be good enough, the committees stopped trying, and members started losing interest in the process fairly quickly. Fire department management then started making mandates for the committees—deadlines, required deliverables, and so forth—and the process then became a burden. Any resemblance to Kaizen or continuous quality improvement disappeared.

Organizations that strongly value customer perceptions as they relate to quality service typically do the following.

1. Think about and talk to their customers (both internal and external).
2. Continually assess their customers' perceptions.
3. Value goodwill versus economic stake.
4. Make amends for poor treatment.
5. Employ a "whatever-it-takes" policy.
6. Redesign systems and kill sacred cows when they obstruct service quality.

Sometimes, when starting out with this process, comprehensive changes must take place. This often occurs when a new fire chief is brought in from another department, or a progressive individual is appointed from within a department or elected in a volunteer department. Change can often be difficult in a fire department, especially when tradition or traditional values are being confronted. Still, to improve quality, significant change must sometimes occur.

Many fire departments are turning to agency accreditation as a measure of quality. Much like consensus standards, criteria are established for a fire department in many different areas. Unlike standards, however, the criteria are not "minimum," but rather could be considered "preferred," and must be met in order to be "accredited." The ambulance industry has a similar accreditation process, and there is now an accreditation process for the position of fire chief as well. (Specific information on these accreditation processes can be found in Appendix C.)

◆ NEED FOR QUALITY MANAGEMENT IN THE FIRE SERVICE

Is quality really needed in the fire service? Is there a need to achieve excellence? If a fire department can extinguish fires, or at least keep them from spreading beyond the building of origin, is there really a problem?

To successfully evaluate the quality of a fire department, managers (or evaluators) must look at products, services, and communications first separately, then how they operate as a whole. Fire suppression services are just part of a fire department's responsibilities. If a fire department provides emergency medical services, these services are certainly more visible because EMS incidents occur about four times more often. Regardless, records are likely kept on the types of fires and types of EMS calls to which the department responds. Public fire/injury prevention education activities are also very visible (because they are meant to be). All of these services—fire suppression, EMS, and public fire/injury prevention education—need to be looked at individually. But they also need to be looked at as a system: Are fires and injuries being prevented, or are fire/injury prevention education programs being run simply for the sake of running them? Are they addressing specific fire or injury problems in the community, based on the fire and EMS call types recorded? Is the department making a difference? Is it the right difference? Can the department prevent more fires or injuries?

Many private fire and EMS departments, such as Rural Metro, do evaluate their services this way. Certainly, they include cost-effectiveness in this review—but every fire department should. Running a fire department *is* running a business, whether or not the bottom line is a profit margin or more funding for additional or improved services. Private fire departments tend to focus more on prevention because the fewer calls they respond to, the fewer expenditures they have. Most municipal fire departments have a mission statement (or some semblance thereof) that says in some way, "to prevent fires," yet focus much more on fire suppression.

Within EMS, another clear example can be shown. Many more lives can be saved with prevention programs aimed at reducing sudden cardiac deaths than can be saved with cardio-pulmonary resuscitation and automatic external defibrillators, yet look at how much the fire service spends (in time and money) on CPR training and equipment. If, nationwide, the fire service were to spend the same amount of time and money on programs like cholesterol screening, blood pressure checks, and heart-safe education, many more people would avoid sudden cardiac arrest than are successfully resuscitated after an arrest occurs (Dyar & Sachs, 1998). Obviously, this is not feasible—people will still have heart attacks, and the fire service needs to be ready to treat them. Still, the concept remains sound. Similarly, if all fire departments were to provide bicycle helmets and promote their use, the incidents of traumatic brain injury would drop significantly (Dyar & Sachs, 1998). This would save millions of health care dollars and rehabilitation costs. Is this important to a fire department manager? Probably not, but it should be. In addition to the positive exposure received from such a program, fewer public dollars spent on health care may mean more dollars available for public safety programs, including those conducted by a fire department. All of this adds to the level of excellence in the department.

Quality is often based on the image of the organization. Again, perception is reality. This is sometimes referred to as the "Disney Model of Customer Service." This model focuses on "moments of truth"—actual interaction with the customer. Fire departments have these moments of truth all the time. Any time the public comes into contact with members of the department—at the grocery store, on the highway as an engine or ambulance passes, at an EMS call for a loved one, as a neighbor's garage burns—is a "moment of truth." Stories in the media can be major moments of truth. These instances can make or break the reputation—perceived quality—of a fire department.

Consider, for example, the major metropolitan fire department's firehouse sex scandal in the late 1990s. Did the events portrayed in the media relate to the quality of service provided to the citizens of the city? Probably not. However, the reputation of the department was significantly tarnished, and the general public may not have viewed the fire department as highly as they had before the stories broke, especially if this was their moment of truth. Similarly, the fire service as a whole, and particularly the Fire Department of New York, gained a solidly heroic reputation after the September 11, 2001 terrorist attacks.

Was anything really done differently that day to deserve such a change in public perception? Again, probably not. However, it was a true moment of truth for us all. It also pointed out that the greatest motivator of all is pride. Starting out with a positive reputation is obviously better than overcoming a negative one, especially if neither has anything actually to do with quality except for that ever-present "perception"—which can mean everything.

Fire, rescue, and EMS personnel across the country received tremendous public support following the heroic actions of responders in New York City, Arlington, Virginia and Shanksville, Pennsylvania during the terrorist attacks of September 11, 2001. *(Photo by Andrea Booher, Federal Emergency Management Agency.)*

Other dimensions of quality that can and should be addressed by fire departments today include the following.

Aesthetics: Appearance, fits, finishes. Example: In EMS, the appearance of personnel; interior appearance of the medic unit.

Conformance: The match with specifications of pre-established standards. Examples: The selection of appropriate mitigation strategies in hazardous materials; the use of the proper PPE in EMS; the enforcement of safe procedures in firefighting (often seen as negative examples in magazine photos).

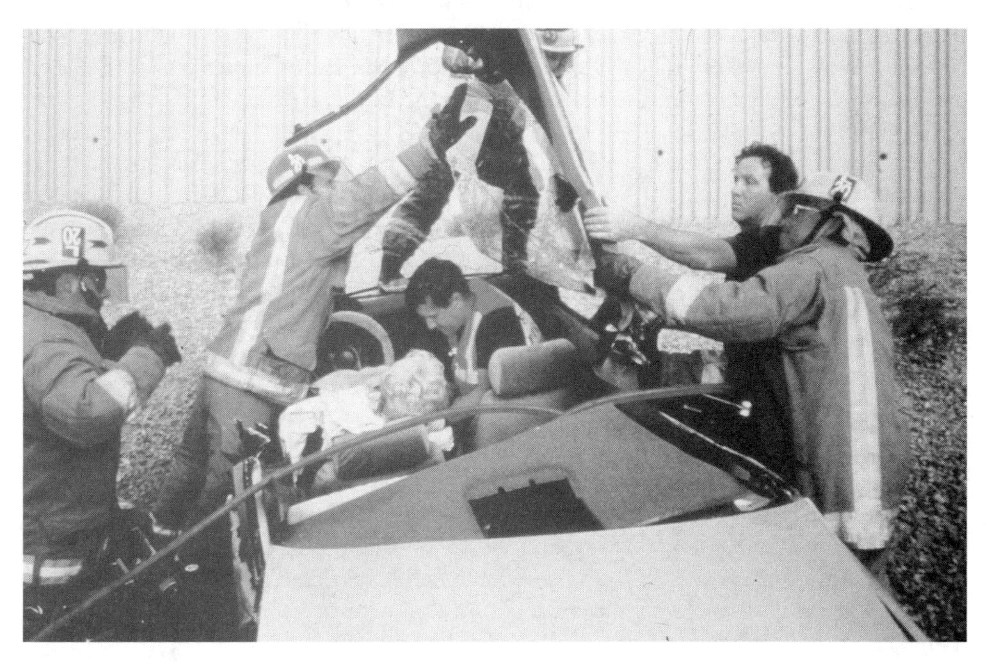

In addition to the obvious risks involved, allowing EMS providers to operate in an area where firefighters are required to wear personal protective equipment often raises questions about professionalism and conformance to standards among the public. *(Photo by Federal Emergency Management Agency, United States Fire Administration.)*

Fire departments are most noted for firefighting operations; performance in this area is critical but is only one part of what fire departments are expected to perform. *(Photo by William Morton.)*

Durability: Product or service life. Example: The efficiency and effectiveness of learning strategies in Public Fire Education (retention of knowledge, skills, etc.).

Features: The secondary characteristics of the product or service. Example: Performing effective salvage operations or taking care of the customer.

Perceived Quality: Reputation. Example: In Public Education, one teacher overhears another discussing the "wonderful" burn prevention program that your fire department delivered. In Rescue, the FEMA US&R task forces.

Performance: The primary operating characteristic of a product or service. Example: The successful completion of tactical assignments during fire suppression.

Reliability: The frequency with which a product or service fails. Examples: A response unit getting lost or breaking down in EMS or suppression; lack of sufficient daytime staffing in a volunteer service resulting in the inability to respond; an inspector forgets to make a follow-up inspection for Code Enforcement.

Preventive maintenance, even during long operations, can keep apparatus running and prevent embarrassment or even undue hazards for fire departments and members. *(Photo by Bryan Day, Bureau of Land Management.)*

Communications/dispatch centers are the first point of contact for fire department customers, even when these centers are managed by a different agency. First impressions are lasting impressions, so the quality of these centers is important to a fire department. *(Photo by Gordon Sachs.)*

Serviceability: The speed, courtesy, and competence of repair. Examples: Ease of acquiring a fire report for insurance purposes for Fire Investigation and Cause Determination; for communications, the speed of answering a 9-1-1 call, the attitude of the call taker/ dispatcher, and the speed with which the call was dispatched.

CHAPTER SUMMARY

To remain a viable public function, fire departments must maintain a constant focus on quality. This focus must not be myopic, reflecting only the point of view of department managers or municipal officials. Nor should it be a tainted view that portrays "what the boss wants to hear." The focus on quality should be based on internal and external customer perceptions and needs, national standards, other fire departments, and other industries. It should be an open, on-going process, and the department's values and vision should reflect that quality is an important aspect of all facets of the department. The focus on quality should be pervasive throughout all divisions and all ranks of the department. Slogans like "Quality is Job #1" and "In Search of Excellence" are not just cute sayings; they are the basis for a cultural change that can make, or break, even the best of fire departments.

References

Davidow, W. H., & Bro, U. (1989). *Total customer service.* Harper and Row.

Deming, W. E. (1986). *Out of crisis.* Cambridge: Massachusetts Institute of Technology, Center for Advanced Engineering Study.

Dyar, J. & Sachs, G. (1998). Is EMS making the right difference? *Fire Chief* 42(5), 42–50.

Evans, B. (2001). Shopping for standards. *Fire Chief* 45(9), 18–19.

Merriam-Webster Online Collegiate Dictionary 2001. [On-line]. Available: http://www.m-w.com

Taguchi, G., & Clausing, D. (1990). Robust quality. *Harvard Business Review* 68(1), 65–75.

Tallon, T. J. (1993). Introducing TQM to the fire department. *American Fire Journal* 45(8), 12–13.

United States Fire Administration, National Fire Academy. (1999). *Advanced leadership issues in EMS.* Washington, DC: U. S. Government Printing Office.

United States Fire Administration, National Fire Academy. (1999). *Executive development.* Washington, DC: U. S. Government Printing Office.

Zeithaml, V., Parasuraman, A., & Leonard, B. (1990). *Delivering quality service.* New York: Macmillan Free Press.

Overview of Quality Management

◆ INTRODUCTION

"Quality Assurance," "Total Quality Management," "Continuous Quality Improvement," are all names for the quality movement that has been around for many years in the business world. As the fire service incorporated emergency medical services (EMS) into their scope of services, the concepts of quality assurance and quality management became familiar terms *in relation to EMS.* The National Highway Transportation and Safety Administration's publication of *A Leadership Guide to Quality Improvement for Emergency Medical Services Systems* in 1997 exemplifies the importance of quality management in EMS. Yet, these principles have not been extended to the fire service as a whole. Services provided by fire departments are no longer restricted to fire fighting. The fire service now responds to multiple community needs including EMS, hazardous materials, confined space rescue, prevention and education, and tactical EMS. The expansion of services involves various job duties for multiservice providers.

As the fire service becomes more and more an all-hazards service, it is critical to make quality management an integral part of the day-to-day operations of the entire organization. In order to make quality management effective, the chief officer and senior leaders must first understand the general concepts of quality management as they relate to the fire service. This chapter will focus on the concepts of three pioneers in the field of quality management and the U.S. government's efforts to promote quality management through the Malcolm Baldrige Award.

◆ CONCEPTS OF QUALITY MANAGEMENT

W. Edwards Deming began his work in Japan in 1950. His philosophy is based on the idea that problems in a production process are due to flaws in the design of the system, not in the motivation or commitment of the workforce. Deming promotes the use

FIGURE 2.1 ◆ Deming's Principles Applied to the Fire Service.

- The fire service can and should be made better.
- Efforts to improve the fire service quality should be continuous.
- Every fire service process yields data and information on how well the process works.
- Data and information are essential to improving quality in the fire service.

of the PDCA cycle (plan, do, check, act): *plan* to implement a policy to improve quality and/or decrease cost of providing services; *do* by implementing the plan; *check* to see if the plan worked; and *act* to either stabilize the improvement or determine what went wrong. PDCA is continuous. The improvements gained in one cycle become the baseline for an improvement target in the next cycle. The principles behind the PDCA cycle related to the fire service are summarized in Figure 2.1.

Deming (1986) also developed "14 points" to transform management practices. Following is a summary of these points applied to the fire service.

1. Create constancy of purpose. A fire service organization's highest priority is to provide the best quality in all services (e.g., fire suppression, hazmat, rescue, EMS, inspection, public education, etc.) offered to its community at the lowest possible cost. The organization must strive to maximize efficiency and effectiveness through constant improvement.

2. Adopt a new philosophy. Everyone in a fire service organization can find ways to improve the system and promote excellence and personal accountability. Pride must be emphasized from recruitment to retirement. The chief officer sets the standard for all workers by his or her behavior.

Mike Staley, in his book *Igniting the Leader Within* (1998), suggests that the pyramid design of fire service leadership that places the chief at the top and members at the bottom be inverted. He states:

> The leader, now on the bottom, serves and supports high-level management, who in turn serves and supports middle management, who in turn serves and supports the members in the trenches every day, doing the job . . . serving the community by providing emergency services. (p. 34)

This philosophy is key to a strong quality management program, but would require a change in the organizational culture of most fire service systems.

3. Stop dependence on inspection to achieve quality. Total reliance on a search for errors, problems, or deficiencies assumes that human performance error or equipment failure is highly likely. There should be a continuous effort to minimize human error and equipment failure. Lasting quality comes from improvements in the system. For example, documenting deficiencies in record keeping related to fire responses does not, by itself, provide a way to make record keeping less error-prone. A quality-driven approach might be the development of a new, clear, and simple record-keeping form.

4. Do not purchase on the basis of price tag alone. A fire service organization must account for the quality of the items it purchases, as well as the cost. If items are purchased on "low bid" alone and do not meet the performance needs of the organization or must be replaced before scheduled, the lack of quality may end up costing more than

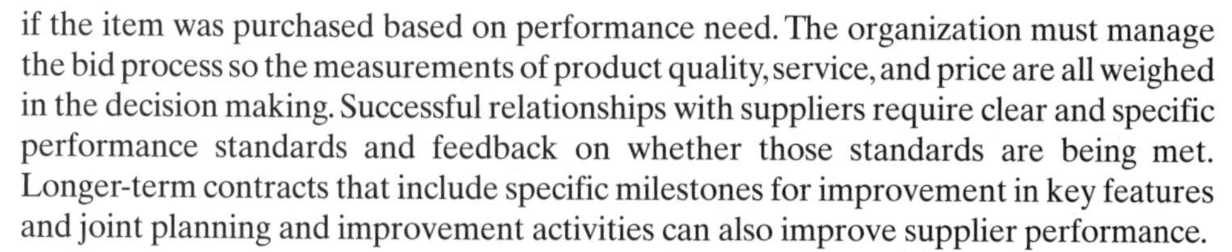

if the item was purchased based on performance need. The organization must manage the bid process so the measurements of product quality, service, and price are all weighed in the decision making. Successful relationships with suppliers require clear and specific performance standards and feedback on whether those standards are being met. Longer-term contracts that include specific milestones for improvement in key features and joint planning and improvement activities can also improve supplier performance.

5. Constantly improve the system of production and service. Quality can be built into all fire organization activities and services and assured by continuous review to identify potential improvements. This requires cooperation between the fire organization that provides the services and the community and members who receive the services. Improving efficiency and effectiveness involves not only meeting present performance targets, but also communicating with the community and members to identify new needed services and performance levels.

6. Institute quality management training on the job. On-the-job quality management training ensures that every member has a thorough understanding of the needs of those who use or pay for the services provided by the fire organization, how to meet those needs with high quality and cost-effectiveness, and how to improve the organization's ability to meet the needs of all customers.

7. Institute effective leadership. Effective leaders know about the work being done and understand the environment and complexities of the job with which their members must contend. Effective leaders create opportunities for members to suggest improvements and act quickly to make needed changes in the processes of the organization. They are concerned with success as much as with failure. They focus on understanding above average performance as well as substandard performance. Effective leaders create opportunities for all members, regardless of level of performance, to interact and suggest methods for improvement.

8. Drive out fear. Errors and failures are opportunities for improvement. Yet, many chief officers believe that the idea of making improvements is an admission that the current way of doing things is flawed, that their management decisions are poor, or that admitting the need for improvement might lead to community distrust or even loss of job security. The fear of identifying problems or needed changes cannot only kill quality management programs, but can also create an unpleasant work environment for all members and place the community at risk. Improved performance occurs only when members believe that they can speak truthfully and that their suggestions will be taken seriously. Both leaders and members must assume that everyone in the fire organization is interested in doing the job to the best of their ability.

9. Break down barriers between departments. Barriers between departments within the organization (or between organizations) are obstacles to effective quality management. Friction or lack of cooperation between organizations or within organizational departments leads to waste, errors, delay, and unnecessary duplication of effort. A lasting quality management program requires teamwork that crosses traditional organizational lines. Managers, members, divisions, and units must share a unified purpose, direction, and commitment to improve the organization.

10. Eliminate member targets for zero defects and new levels of productivity. The problem with zero defects for members is that it puts the burden for quality on member performance instead of poor system design. Quality management requires the organization to focus on improving work processes. Improved work processes increase service quality, productivity, and efficiency. The following is a very simplified example of what is meant by this concept.

A call is received that there is smoke coming from a top story window of a two-story dwelling. The structure is 25% involved on the second floor when the first-in company arrives. The building ends up as a total loss. When the incident is reviewed, the following factors were noted.

- The response time of the first-in company was 12 minutes (for what should have been a 6-minute response time).
- The Incident Commander was a newly promoted officer; this was his first major incident.
- There was a delay in getting water to the fire.

Often the initial response is to assume these problems resulted from a lack of member performance (i.e., the engine company got lost, the Incident Commander lacked adequate knowledge in ICS or fire suppression tactics, etc.). Rather than assume most problems are member related, the organization must begin to look at system factors. For instance, was there a traffic flow problem, and has this occurred in the past? If so, the organization might need to look at alternate routes or traffic management approaches such as remote signal control devices. If a volunteer department was involved, was there a daytime staffing problem? Was there a maintenance problem with the apparatus or hydrant? If so, this might indicate the need to review the preventive maintenance program of the department.

11. Eliminate management by numbers and objective. Deming suggests that work standards tied to incentive pay are inappropriate because they burn out the members in the long run. Implementing a team effort to increase quality by improving processes will lead to improvement in outcome measures. This improvement will bring increased profits and savings that can then be translated to higher salaries or better benefits. Although specifically tied to a product, this concept can be applied to a career or volunteer fire service organization. For example, the organization sets a goal to install smoke detectors in all low-income areas in the community in order to reduce fire injuries. If this were accomplished, one could say that the department was 100% effective in meeting their objective. However, if personnel were not involved in determining ways to improve the community's knowledge and understanding of the need to change batteries on a regular basis and in promoting fire safety education, the objective has not been met at all. Establishing a team that includes managers, line personnel of that heritage, the public education division, and community members to identify ways of providing effective education improves the process of reducing fire injuries. In addition, the organization would need to follow through to assure that all members installing the detectors were given the important points to teach.

12. Remove barriers to pride in workmanship. Members are the most important component of the fire service. A fire organization cannot function properly without members who are proud of their work and who feel respected both as individuals and professionals. Managers must make sure that job responsibilities and performance standards are clearly understood from the time of hiring, build strong relationships with members through open communication and respect, and provide the best equipment, supplies, and information possible.

13. Institute a strong program of education and self-improvement. The fire organization needs not just good people; it needs people who are growing through education and life experiences. Both leaders and members must continue to experience new learning and growth if the organization is to provide the best quality services to the community. Chief officers must continue to learn if they are to identify new community needs and to stay abreast of new developments in technology and management practices. Strong chief officers view education of their members as not only necessary to provide new services or use new equipment, but also as the development of competent and qualified replacements to provide the future leadership of the fire service.

Developing this learning organization may include the implementation of a mentoring program. Leaders must be willing to share not only what they have learned from books, but also what they have learned through their personal successes and failures. Failures are only true failures if nothing was learned from the experience. This approach may also require a change in organizational culture, for it means that the leaders must admit to the followers that they are only human and have made mistakes.

14. Put everyone to work to accomplish the transformation. Effective quality management programs go beyond emphasizing one or two efforts or areas to improve performance. Every activity, process, and job in the fire service can be improved in some way. The first step is to give everyone in the organization an opportunity to understand the quality management program and his or her role within that program. The second step is to develop improvement teams that include representation from all organizational departments to provide interdepartmental communication and information exchange.

Joseph Juran based his approach to quality management on the philosophy that a strong interdependency exists among all of the operations within an organization. His approach to quality management involves three aspects: quality planning, quality control, and quality improvement (Juran & Gryna, 1988). *Quality planning* is the process of identifying and understanding what the customer needs, and then designing all aspects of the system to meet those needs reliably. *Quality control* is used to constantly monitor performance for compliance with the standards that were designed during the planning stage. If performance standards are not met, corrective action is implemented. *Quality improvement* occurs when previously unmet levels of performance are achieved. Figure 2.2 summarizes Juran's trilogy applied to the fire service.

Juran proposed the idea of the "vital few and the useful many" to help prioritize which quality management projects to undertake. With the multitude of services provided by the various divisions within a fire service organization, the list of possible ideas for improvement will be very lengthy. Because resources to implement new ideas are limited, chief officers must choose the vital few projects that have the greatest impact on improving the organization's ability to meet customer needs. Determining which projects to select involves considering the potential impact on meeting customer needs, cutting waste, and attaining the resources required to implement the project.

Juran also suggests instituting a senior staff leadership group responsible for designing the overall strategy for all three aspects of the program (planning, control, and

- *Quality Planning*—Design operations based on meeting customer needs.
- *Quality Control*—Continuously monitor how the organization is maintaining established performance levels, and take corrective action when needed.
- *Quality Improvement*—Create integrated quality improvement teams to plan, test, and implement new methods to reach higher levels of performance.

FIGURE 2.2 ◆ Juran Trilogy Applied to the Fire Service.

improvement) because quality management is as important as budgeting, human resource management, purchasing, and training. This senior level group can provide integration of quality management into every aspect of fire service operations.

Philip Crosby (1979) coined the phrase "Quality is free." He suggests that the lack of quality is costly to an organization in money spent on doing things wrong, over, or inefficiently. Spending money to reduce waste or improve efficiency saves money in the long run.

Crosby focuses his approach to quality management on the design phase. He proposes that organizations think of how processes can be designed or redesigned to reduce errors. He also suggests a step-by-step education program for the entire workforce about quality principles, extensive measurement to document system failures, and formal programs to redesign unsatisfactory processes.

THE MALCOLM BALDRIGE AWARD

In the early and mid-1980s, many industry and government leaders saw that a renewed emphasis on quality was a necessity for U.S. companies doing business in the expanding and more competitive world market. But many U.S. businesses either did not believe quality mattered for them or did not know where to begin. The Baldrige Award was envisioned as a standard of excellence that would help U.S. organizations achieve world-class quality.

The award was created by Public Law 100–107 in 1987 and named for Malcolm Baldrige, who served as secretary of commerce from 1981 until his death in 1987. The U.S. Commerce Department's National Institute of Standards and Technology (NIST) manages the award program in close cooperation with the private sector. In 1999, the award was extended to education and health care.

Applications for the award are judged on seven criteria: leadership, strategic planning, customer satisfaction, information and analysis, human resource development, process management, and system results (Figure 2.3).

Additional information on the Malcolm Baldrige Award can be found on the National Institute of Standards and Technology homepage at http://www.quality.nist.gov.

FIGURE 2.3 ◆ Baldrige Categories.

- Leadership
- Customer Satisfaction
- Human Resource Development
- Strategic Planning
- Information and Analysis
- Process Management
- System Results

CHAPTER SUMMARY

A strong quality management program is essential for any organization to satisfy community needs in a cost effective manner and to remain competitive. The principles and methods of Deming, Juran, and Crosby provide the basis for most quality management efforts. The Baldrige criteria are used by thousands of organizations as a guide for self-assessment and training and as a tool to develop performance and business processes. Using the criteria can lead to better member relations, higher productivity, greater customer satisfaction, and improved cost-effectiveness.

The remainder of this book addresses each of the seven Baldrige criteria using the principles and methods of Deming, Juran, and Crosby and applies them to fire service organizations. Although reading how the work of quality management experts can be applied to the fire service is a start, it will be the chief officer's personal and professional commitment to apply these principles that provides the most important contribution for success.

Management Transformation Checklist

[ ] **Create constancy of purpose.**
- [ ] Review policies and procedures to maximize efficiency and effectiveness.

[ ] **Adopt a new philosophy.**
- [ ] Set the standard for all workers.
- [ ] Support management and members.

[ ] **Stop dependence on inspection to achieve quality.**
- [ ] Develop methods to minimize human error and equipment failure instead of just tracking problems.

[ ] **Do not purchase on the basis of price tag alone.**
- [ ] Develop clear and specific performance standards for equipment and supplies.
- [ ] Develop feedback mechanism from members to determine field performance.
- [ ] Manage the bid process so the measurements of product quality, service, and price are all weighed in the decision making.

[ ] **Constantly improve the system of production and service.**
- [ ] Communicate with the community to identify new needed services and performance levels.
- [ ] Communicate with members to identify needs to improve performance levels.

[ ] **Institute quality management training on the job.**
- [ ] Educate every member in:
  - the needs of those who use the services.
  - how to meet those needs with high quality and cost-effectiveness.
  - how to improve the organization's ability to meet the needs of all customers.

[ ] **Institute effective leadership.**
  [ ] Create opportunities for members to suggest improvements and act quickly to make needed changes in the processes.
  [ ] Focus on understanding above average performance as well as substandard performance.

[ ] **Drive out fear.**
  [ ] Take members' suggestions seriously.
  [ ] Build an atmosphere of trust to encourage members to speak truthfully.

[ ] **Break down barriers between departments.**
  [ ] Share communications among departments.
  [ ] Institute interdepartmental teams to problem solve.

[ ] **Eliminate member targets for zero defects and new levels of productivity.**
  [ ] Focus on improving work processes.

[ ] **Eliminate management by numbers and objective.**
  [ ] Implement a team effort to increase quality by improving processes.

[ ] **Remove barriers to pride in workmanship.**
  [ ] Ensure job responsibilities and performance standards are clearly understood from the time of hiring.
  [ ] Build strong relationships with members through open communication and respect.
  [ ] Provide the best equipment, supplies, and information possible.

[ ] **Institute a strong program of education and self-improvement.**
  [ ] Provide educational opportunities for management and members.
  [ ] Provide educational programs that develop competent and qualified replacements to provide the future leadership of the fire service.
  [ ] Implement a mentoring program.

[ ] **Put everyone to work to accomplish the transformation.**
  [ ] Give everyone in the organization an opportunity to understand the quality management program and his or her role within that program.
  [ ] Institute improvement teams that include representation from all organizational departments to provide interdepartmental communication and information exchange.

References

Crosby, P. B. (1979). *Quality is free: the art of making quality certain.* New York: McGraw-Hill.

Deming, W. E. (1986). *Out of crisis.* Cambridge: Massachusetts Institute of Technology, Center for Advanced Engineering Study.

Juran, J. M., & Gryna, F. M. (1988). *Juran's quality control handbook.* New York: McGraw-Hill.

National Highway and Safety Administration. (1997). *A leadership guide to quality improvement for emergency medical services.* Washington, DC: U. S. Government Printing Office.

Staley, M. (1998). *Igniting the leader within.* Saddle Brook, NJ: Fire Engineering.

The Chief Fire Officer's Role in Quality Management

◆ INTRODUCTION

This chapter focuses on the first Baldrige category of leadership. The Baldrige leadership objectives as they relate to the fire service are summarized in Figure 3.1.

The chief officer's role in developing a strong quality management program begins with the creation of a personal and organizational focus on the needs of internal and external customers. Words are not enough. The organization's vision, mission, values, goals, and expectations must reflect a commitment to quality services and performance excellence and the chief officer must demonstrate personal commitment through action. The chief and senior officers must work with the managers of the organization to create a strong customer focus and define the vision statement, mission statement, values, operational objectives, and long-term expectations in clear and understandable terms.

Developing a quality management program that results in continuing higher levels of service requires a strategic plan. The strategic plan should:

- identify clear goals that define expected outcomes of the quality management effort
- be based on facts and use indicators to measure progress
- include systematic cycles of planning, implementation, and evaluation
- concentrate on key processes rather than outcome numbers
- focus on customers, both internal and external

Chapter 4 addresses the strategic planning process in detail.

◆ CUSTOMERS

Customers may be defined as *external* or *internal,* depending on how they relate to the fire organization. External customers include all entities that interact with the organization that are outside the actual operation of the organization. External customers

FIGURE 3.1 ◆ Summary of Baldrige Leadership Objectives.

- Chief officer or senior level chief should lead the quality management effort.
- Educate leadership and management in quality management theories, strategies, and benefits.
- Clarify organizational vision, mission, values, and goals.
- Initiate strategic planning.
- Set leadership/management standards, tasks, and procedures.
- Develop policies and actions for community involvement.

include entities such as local, state, and federal governmental agencies that regulate aspects of service, the community, law enforcement, patients served by the EMS component and their families, and insurance companies and third-party payers that pay for services. Internal customers are those who are involved in or with the operation of the organization. Internal customers include the organization's employees and volunteers, union, the agencies that interact with the organization to provide integrated services such as disaster response, and hospitals, planning committees, and other health care providers that work with the organization to provide health care to citizens of the community. All services provided by a fire organization have external and internal customers, and these customers may be different for different service areas (e.g., EMS, disaster planning, technical rescue, hazmat, inspection, training, public education).

Focusing on customers begins with identifying those individuals and entities. This can be done by having each division of the organization, as well as management, develop a list of both their internal and external customers. It is important that this identification process involve each division and not just upper management. It is the personnel in these divisions who interact on a daily basis with the community and who are often aware of many more customers that upper management would fail to identify. These lists can then be combined for an overall list, as many entries will be repetitive (e.g., OSHA, NFPA, EPA, city government, specific groups in the community, members, etc.).

After the list of customers is developed, their needs and expectations should be identified. It is important to remember that identifying customer needs should include the needs of internal as well as external customers. This can be accomplished in two ways.

Surveys can provide direct and measurable information about which services and parts of those services most affect customer satisfaction, both internal and external. However, for survey information to be relevant, the questions must be written in a form that will yield responses that are useful.

Surveys are only as good as the issuing organization wants them to be, and certainly can be written to obtain the answers the organization *wants* rather than the answers it *needs.* For example, the fire service typically receives very high ratings from the public as a result of the "white hat" or "hero" perception. This is especially true after the September 11, 2001 terrorist attacks. These exemplary ratings mean the public has a high regard for the fire service, but it does not necessarily mean every fire service organization is a good quality organization. Care must be taken to write questions that will yield useful information, not just information that can be used to justify the organization's present condition. It may be worth the time and expense to develop surveys in conjunction with an outside, objective, trained survey writer. Specific survey questions must be developed for each particular customer group from whom you are seeking in-

Surveys are one good way to identify what your internal and external customers expect from their fire department. *(Photo by Gerry Suftko, Mesa Fire Department.)*

formation. Remember that members are internal customers, but they deal with external customers every day. A separate survey to solicit member input can be an invaluable source of information about both internal and external customer needs.

Focus groups are composed of representatives of customer groups who identify their groups' needs in face-to-face discussion. Focus groups often yield more detailed information regarding needs than surveys, but the results of these discussions may be hard to quantify for analysis. Conducting a focus group requires planning.

Topics should be identified before the groups meet. The number of participants should be limited to no more than 10, and the meeting should last no more than two hours. A note-taker should be present. The manager conducting the focus group should be a skillful facilitator. His or her role is to encourage sharing of ideas and keep the group on track, not to direct or lead the discussion. He or she must remember that the purpose of the focus group is to identify what the customers believe are their needs, not to direct or reinforce what the organization believes are the customers' needs. Figure 3.2 summarizes the steps for organizing a focus group.

Well-written surveys and well-conducted focus groups can identify both external customer expectations regarding operational services and expectations regarding timeliness of response, ease of access to the system, and level of courtesy and caring demonstrated by personnel. Internal customers can provide leaders with information that leads to improved training programs, management and human resource issues, and job safety concerns.

Obviously, every need customers identify cannot be met. Some may result from an individual participant's desire. Some may be minor and the result of one specific

FIGURE 3.2 ◆ Focus Groups.

- 5–10 participants
- meet for 1–2 hours
- discuss pre-identified topics
- encourage sharing of ideas
- note-taker records information
- conduct additional focus groups on same topic until information becomes repetitive

experience with the organization. Some may be cost- or staff-prohibitive, or subject to legal restrictions. Some needs may already be met through the services provided by another agency. The leadership of the organization must take the input collected from the customer contacts and develop a list of key customer requirements. These key customer requirements must then be prioritized and should form the basis for the organization's mission, vision, and values statements, as well as its strategic planning goals and objectives. This is an important leadership responsibility. These documents must reflect the viewpoints of not only the customers of the local fire organization, but also the viewpoints of regional and national fire service organizations.

◆ VISION, MISSION, VALUES, AND GOALS

The quality of a fire organization hinges on its vision, mission, values, and goals (Figure 3.3). Leaders who lead without a vision of what their organization is to become doom that organization to function based only on tradition. Without a vision of the organization's future, leaders are reduced to keeping things the way they have always been, guided by the archaic defense, "If it ain't broke, don't fix it."

True leaders do things differently. They live by the saying, "If it ain't broke, you're not looking in the right place." They realize there is always room for improvement and believe that no one has ever done anything so well that it cannot be done better. A vision is a reality that does not yet exist.

The *vision statement* declares where the organization wants to be in the future and serves as a major focal point of strategic planning. The vision is the target toward which the chief officer aims his or her resources and energy and provides a focus despite obstacles such as negative attitudes of superiors, peers, members, or practical difficulties. When shared by members, the vision can keep the entire organization moving forward in the face of adverse circumstances because they understand what the chief officer is trying to accomplish and what the organization stands for.

FIGURE 3.3 ◆ Vision, Mission, Values, and Goals.

Vision: Future direction of the organization
Mission: Purpose of the organization
Values: Beliefs and principles
Goals: Proposed accomplishments

Ideally, the vision statement should be developed with member input. However, it is possible to obtain member buy-in without their actual involvement in the development process if they understand the vision and are involved in the implementation process. The chief officer must communicate the vision to others for it to become shared. He or she must first act in a manner consistent with the vision and not send mixed signals by saying one thing and doing another. After the vision has been explained, members must decide whether they want to be a part of it. They cannot be forced to support the vision over a long period of time without considerable cost to the organization. The day has nearly passed when autocratic chief officers can succeed over the long term. The cost of using this approach is too high in terms of inferior output resulting in poor quality, lost member loyalty and support, and money.

Most members are motivated by achievement, recognition, self-esteem, and the sense of having lived up to their ideals. The chief officer must connect with these needs and respond to the ideas that rise out of the organization by involving the members in deciding how to improve and achieve the vision and recognizing and rewarding them for their contributions. Even if the chief officer creates the vision solely, a shared vision can still result if members are allowed to influence the implementation of that vision. Simply put, involvement creates ownership. Implementation approaches are often best decided with input from those who must actually carry out the implementation steps. And who better knows how changes may affect day-to-day operations? In order to remain applicable, vision statements also must be reviewed and revised as changes in community vision, department mission, and regional needs occur.

Sam Walton changed the face of American retailing when he created Wal-Mart. His vision was far-reaching, but simple, and shared by all his associates. From humble beginnings in Arkansas, he built a retail empire recognized worldwide. As a result, he had a profound effect on the practices of other retail chains and became one of the richest men in this country. His vision can be found in Figure 3.4. Figure 3.5 is an example of a vision statement related to emergency services.

The *mission statement* identifies the purpose of the organization. It should describe all the essential components of the organization, such as the customers, geographic service area, major services provided, economic goals, and organizational strengths. Developing a good mission statement is time consuming and addresses several elements.

- **Basic Elements**

  What services the organization provides. The definition must expand past just services provided and focus on customers' needs.

  Who receives the services. An organization can identify its customers in many ways: geographically, revenue generation, businesses, and so forth. Clarity about customers allows the organization to focus its resources.

> We're all working together; that's the secret. And we'll lower the cost of living for everyone, not just in America, but we'll give the world an opportunity to see what it's like to save and have a better lifestyle, a better life for all. We're proud of what we've accomplished; we've just begun.
>
> Sam Walton
> (1918–1992)

FIGURE 3.4 ◆ Vision for Wal-Mart.

FIGURE 3.5 ◆ EMS Vision Statement.

> Emergency medical services (EMS) of the future will be community-based health management that is fully integrated with the overall health care system. It will have the ability to identify and modify illness and injury risks, provide acute illness and injury care and follow-up, and contribute to the treatment of chronic conditions and community health monitoring. This new entity will be developed from redistribution of existing health care resources, and will be integrated with other health care providers and public health and public safety agencies. It will improve community health and result in more appropriate use of acute health care resources. EMS will remain the public's emergency medical safety net.
>
> *EMS Agenda for the Future*
> National Highway Traffic Safety Administration

How services are delivered. This includes services such as EMS, hazmat, inspections, public education, prevention programs, public gathering standbys, and others.

Why the organization exists. This is usually a simple statement such as "To save lives and protect property."

- **Driving Forces**

  The driving forces of the organization should be identified and prioritized by a strategic planning team. Most major strategic decisions involve the allocation of resources. If resources are inadequate, the ranking of the forces that have been identified can determine how resources will be allocated or which direction the organization will go.
- **Organizational Attributes**

  The organization should identify the aspects and services that only it provides and those that set them apart from competitors. Distinct attributes may include such things as specialty services or specific prevention programs.

After all of these elements have been identified, they should be integrated into the organization's mission statement. The mission statement should be brief and clearly identify the organization's basic service. The mission statement helps the organization develop its course of action and provides a guide for making routine, day-to-day decisions. Once written, it is critical that all members of the organization know and understand it. Figures 3.6 and 3.7 feature examples of organizational mission statements.

After the overall mission statement has been developed, mission statements that are more specific should be developed for each division within the organization. Division mission statements should be more focused and more limited, but must be de-

FIGURE 3.6 ◆ Organizational Mission Statement.

> To provide high quality apparatus manufactured to the customer's satisfaction and to maintain a high quality service department for before and after delivery service.
>
> New Lexington Fire Equipment Co., Inc.

FIGURE 3.7 ◆ Organizational Mission Statement.

The International Association of Fire Fighters is dedicated to the following objectives: To organize all fire fighters and emergency medical or rescue workers; to secure just compensation for their services and equitable settlement of their grievances; to promote as safe and healthy a working environment for the fire fighters and emergency medical or rescue workers as is possible through modern technology; to promote the establishment of just and reasonable working conditions; to place the members of the Association on a higher plane of skill and efficiency; to promote harmonious relations between fire fighters and their employers; to encourage the formation of local unions, state and provincial associations and joint councils; to encourage the formation of sick and death benefit funds; to promote the research and treatment of burns and other related health problems common to fire fighters and emergency medical or rescue workers; to encourage the establishment of schools of instruction for imparting knowledge of modern and improved methods of fire fighting and prevention as well as emergency medical or rescue technology; and to cultivate friendship and fellowship among its members.

Constitution Preamble
International Association of Fire Fighters

FIGURE 3.8 ◆ Division Mission Statement.

The purpose of the Occupational Health and Safety Department is to develop knowledge within the fire service so fire fighters, paramedics and EMTs can recognize and control the safety and health hazards associated with the profession. To assist in the achievement of that goal, the Department offers a comprehensive array of services addressing occupational health and safety. Additionally, the Department is responsible for the Occupational Medicine Residency Program, the IAFF Standing Committee on Occupational Health and Safety, the Redmond Foundation, the IAFF Representatives on various Standards Development Committees, the IAFF Cancer Study Program, the PSOB Program and the Ad Hoc Committee on Labor and Employee Assistance Programs.

Health and Safety Department
International Association of Fire Fighters

rived from the overall mission statement. Figure 3.8 provides an example of a division mission/purpose statement.

The *values statement* identifies the basic beliefs and principles regarding how all of the organization's members will work together. The values statement covers issues such as fairness, honesty, commitment, dependability, and expectations. The values statement does not need to be a lengthy list of expected behaviors. Sam Walton made people feel welcome and important. He passed this on to his Wal-Mart associates. When he visited his stores, he asked the associates to make a pledge: "I want you to promise that whenever you come within 10 feet of a customer, you will look him in the eye, greet him, and ask if you can help him." This pledge is what Wal-Mart now calls its "10-foot attitude," and embodies their values statement. Figure 3.9 features the values statement found on the business cards for the Clackamas County Fire District #1 (Milwaukie, OR).

FIGURE 3.9 ◆ Values Statement.

Do our best. ◆ Let honesty, integrity, courage and personal accountability guide our actions. ◆ Trust each other. ◆ Treat others like we want to be treated. ◆ Leave it better than we found it. ◆ Communicate openly and share knowledge. ◆ Encourage and respect contributions from everyone. ◆ Cultivate the potential for excellence and leadership in all members. ◆ Measure our success by the satisfaction of our customers.

Clackamas County Fire District #1
Milwaukie, OR

FIGURE 3.10 ◆ Operational Goals.

- The identification of the safety and health needs and hazards confronting fire fighters through policies established by Convention and/or IAFF Executive Board; directives from the IAFF General President; computerized surveys; and communications/correspondence with state/provincial associations and local affiliates, liaison with government and non-government organizations, the scientific and medical literature, and through critiques of fire and emergency activities throughout the US and Canada.
- The development of educational materials, including manuals, videos and audiovisual presentations, which specifically address fire fighter occupational safety and health. The preparation of articles for the *International FireFighter* and the *IAFF Leader*, legislative testimony, and emergency mailings to local affiliates on critical safety and health issues. In addition, the development of curriculum for use in regional seminars.
- The maintenance of technical information and data and the dissemination of information to IAFF affiliates, upon their request, for specific safety and health related material.
- The ability to develop and disseminate knowledge by way of direct contact with fire fighters through educational seminars developed by the Department and conducted throughout the United States and Canada in conjunction with the IAFF Regional Seminar Program.

Health and Safety Department
International Association of Fire Fighters

Goals are the proposed accomplishments of the organization. Operational goals and objectives are defined in the strategic planning process and provide day-to-day direction for progress. Figure 3.10 includes some of the operational goals of the Occupational Health and Safety Department of the International Association of Fire Fighters (IAFF).

◆ EMPOWERING THE WORKFORCE

Members perform better and strive harder when they feel personally invested in their work. All members of the organization must feel empowered to make an impact on the quality of services. Through careful planning and transition, the chief officer and managers can maintain authority and responsibility and increase autonomy of input from line personnel.

This transition requires the creation of new working relationships. Training in team dynamics and problem solving can help provide personnel with the skills needed to make these new working relationships succeed. For example, teams of EMS providers can be formed to identify ways to improve the quality of patient care, line personnel can identify apparatus requirements, or integrated performance teams composed of members from multiple divisions that include dispatchers, fleet maintenance, and data collection personnel (as well as line personnel and managers) can focus on ways the entire fire suppression response can be modified to better meet customer needs. At the state or regional level, councils and advisory committees can serve a similar function by providing a forum for developing leadership expertise and consensus on quality management direction.

Managers can be a strong motivating force between the chief officer and line personnel, as they often serve as the connection between senior leadership and the workforce. Therefore, early manager buy-in to quality management activities is critical. The chief officer must create opportunities for managers to develop and improve their quality management skills. Their quality management roles and responsibilities must be clarified and include more quality improvement team facilitation and less inspection or supervision. Increased communication among all organizational levels and divisions should be encouraged and managers should participate in frequent quality, financial, and strategic performance reviews.

A strong quality management program requires that managers receive the time, incentive, and support to participate as both instructors and learners in quality management educational activities, to meet frequently with internal and external customers, to mentor new members in quality management values and policies, and to serve as facilitators of quality improvement teams. Manager activities should be evaluated periodically for desired results. If results do not meet organizational goals, the chief officer and senior leaders need to revise the activities.

◆ LEADERSHIP ACCOUNTABILITY

All members of the organization's leadership system, from the chief officer to the lowest level manager, should assess how well they "walk the talk" of quality management. Regardless of how they perceive their leadership style, the perceptions of the internal customer (member) dictate the amount of support (or non-support) they will receive when seeking organizational changes. This means seeking feedback from line personnel on their leadership performance and how it might be improved. Feedback from line personnel for all levels of leadership and management should be sought, and lower level managers should provide feedback to senior level officers and the fire chief. This is a radical departure from traditional fire organization philosophy.

For feedback to be productive, three things must occur. First, the form used to obtain feedback must be performance-centered. Second, all managers (including the chief) must be open to constructive criticism and view that input as performance related and not personal. Third, there can be no reprisals (direct or indirect) attached to the feedback.

As with customer surveys, careful thought must go into the development of the questions. Figure 3.11 provides some examples of performance-related questions that might be used to seek feedback regarding leadership accountability. This can help prevent "emotional" responses from both front-line personnel and management. Questions

FIGURE 3.11 Leadership Accountability Questions.

- Select the most common method by which you receive communication related to organizational changes.
  - (a) face-to-face from a chief officer
  - (b) written memo
  - (c) e-mail
  - (d) hear from another member
- Rate the explanation of communication related to organizational changes.
  - (a) a comprehensive explanation is provided, with a contact person for questions
  - (b) a partial explanation is provided
  - (c) no explanation is provided
- How are you made aware that a potential change is being analyzed?
  - (a) written or verbal request for member volunteers to serve on committee
  - (b) update memo, written or e-mail
  - (c) face-to-face update from company or chief officer
  - (d) not aware change was coming until memo with effective date

Identify at least one way that communication between chief officers and members can be improved.

____

____

need to relate to quality management policies and issues and should require suggestions to improve performance in each area identified as a problem. It is easy to point out perceived problems, but when those problems must be accompanied by a suggestion for improvement, personnel are far more likely to consider the performance and not the person.

In the early stages, both managers and line personnel will no doubt be apprehensive about the process. Feedback may be less objective than desired. Use of anonymous feedback may help. Eventually, anonymity may not be necessary as fear of reprisal lessens.

COMMUNITY INVOLVEMENT

Fire service organizations are part of the communities they serve and can contribute to community well-being in the same way every citizen is expected to contribute. The chief officer and senior leadership can set high personal and organizational standards for ethical conduct in business and work practices by allowing public accountability and disclosure of performance information. The fire leadership must also assure that their organization and members continually demonstrate professional behavior and values.

Fire organizations also have a responsibility to work with other public safety, health care, and private entities that are involved in the overall system. Actions such

Career and volunteer firefighters across the nation, through the "Fill the Boot" campaign initiated by the International Association of Fire Fighters, are key players in raising funds for the Muscular Dystrophy Association (MDA) each year.

as participating in community-wide planning for EMS services, disaster planning, and fire prevention and inspection programs build strong links with these entities and the community at large. Liaisons with outside entities help maintain a customer focus, as well as serving as a source to identify customer needs. Some departments have commissions made up of citizens to provide quality feedback to the department. Fire organizations can also participate in state and national reporting systems such as the United States Fire Administration's National Fire Incident Reporting System (NFIRS) that allow comparison and benchmarking between systems.

Fire organizations should take a leadership role in public education related to both fire and injury prevention, environmental protection, and other community-wide issues. Educational activities such as fire safety and injury prevention classes, system access training, and pre-arrival emergency care both educate the public and develop goodwill and support within the community. The degree of community involvement for a fire organization depends on the resources available, but all fire organizations can make some contributions either alone or in conjunction with larger organizations to maximize resources.

The chief officer can also support on-the-job or after-hours involvement in organized community programs such as "Jerry's Kids," United Way, blood drives, scouting, and so forth. Participation in these kinds of activities maintains the organization's grassroots links with the community and encourages member leadership. The Volunteer Fireman's Insurance Service (VFIS) book, *Building Blocks,* is a thorough, how-to manual containing instructions and examples for conducting successful community outreach.

CHAPTER SUMMARY

Developing a good quality management program begins with the chief officer. He or she must have a vision for the future of the organization that is shared by members, oversee the development or update of a relevant mission statement, take steps to identify what both internal and external customers need as they see it, and foster strong community relations. This may require a major change in organizational culture; the traditional "trickle down" of policy and procedures needs to be altered to input and feedback coming from the bottom up. Leadership must be open to discovering ways to improve services and improve their leadership abilities. By operating day to day in a manner consistent with a team approach and open communication, the chief officer will "walk the talk" and reap the benefits of improved quality of services and customer satisfaction.

Leadership Checklist

[ ] **Identify Customers.**

- [ ] Have management and each division develop a list of the groups with whom they interact on a regular basis.

External

____________________ ____________________

____________________ ____________________

____________________ ____________________

Internal

____________________ ____________________

____________________ ____________________

____________________ ____________________

- [ ] Review and collate each list to prepare the final list.

[ ] **Identify each customer's needs and expectations.**

- [ ] Identify both current and future needs.

Customer Needs Worksheet

Need	**Approach(es) to Meeting the Need**

*Consider using surveys and focus groups to obtain information.

[ ] **Review, revise, and/or develop the vision, mission, values, and goals.**

- [ ] Use a team composed of members from all divisions and ranks to provide input.
- [ ] *Vision statement*
  Where does the organization want to be in the future?
- [ ] *Mission statement*
  What is the purpose of the organization?

What services does the organization provide? (Focus on customer needs.)	
Who receives the services?	
How are services delivered?	
Why does the organization exist?	
What are the driving forces? (These should be prioritized to aid in decisions involving allocation of resources.)	
What services are provided only by the organization? (Organizational attributes)	

  - [ ] *Values statement*
    What are the basic principles of how the organization members work?
  - [ ] *Goals*
    What are the proposed accomplishments of the organization?

- [ ] **Empower the Workforce.**
  - [ ] Use work teams.
    - [ ] Assure members are trained in team dynamics and problem-solving techniques.
    - [ ] Assure appropriate team composition.
    - [ ] Assure managers function as facilitators rather than supervisors.
    - [ ] Evaluate manager performance.
  - [ ] Increase communication at all organizational levels.
  - [ ] Assure adequate time for manager/team training and functions.

- [ ] **Leadership Accountability.**
  - [ ] Seek feedback from line personnel on leadership performance.
    - [ ] Develop feedback mechanism that is performance-centered.
    - [ ] Be open to constructive criticism.
    - [ ] Assure no reprisals are attached to feedback.
    - [ ] Use anonymous feedback.

- [ ] **Community Involvement.**
  - [ ] Disaster planning
  - [ ] EMS service
  - [ ] Fire prevention and inspection programs
  - [ ] Fire safety and injury prevention public education programs
  - [ ] Pre-arrival emergency care education programs
  - [ ] Organized community programs (e.g., United Way, blood drives)

Activities

1. For each of the categories discussed in this chapter, identify the changes you, as a leader, would have to make personally or within the organization to build the foundation for a good quality management program.
2. Review your current vision and mission statements. When were they written? When were they last updated? Did they include input from all levels of the organization? Do they reflect the current vision and mission of the department? What areas need to be addressed? Rewrite these statements to reflect what you believe they should communicate.
3. Identify the current needs of your customers. Next to each need, write the approach(es) your organization is now using to meet that need. Then list the needs that you foresee your customers having in the next five to ten years. Next to each need, write the potential approach(es) your organization could take to meet that need. You may select just one area of service for this exercise.
4. Compare the vision and mission statements you rewrote in Activity 2 with your current and future needs lists from Activity 3. Are they compatible? How would you change your vision and mission statements to accurately reflect the needs and approaches you identified?

References

About Wal-Mart. (2001). [On-line]. Available: http://www.walmart.com/cservice/aw_story.gsp

International Association of Fire Fighters. (2001). [On-line]. Available: http://www.iaff.org/iaff/Health_Safety/health_safety.html

National Highway and Safety Administration. (1997). *A leadership guide to quality improvement for emergency medical services.* Washington, DC: U.S. Government Printing Office.

New Lexington Fire Equipment Co., Inc. (2001). [On-line]. Available: http://www.newlexfire.com/index.htm

Sparber, P., & Fernicola Suhr, K. (1997). *Building blocks.* York, PA: VFIS.

United States Fire Administration, National Fire Academy. (1999). *Advanced leadership issues in EMS.* Washington, DC: U.S. Government Printing Office.

Strategic Quality Planning

CHAPTER

◆ INTRODUCTION

It is sometimes said that Christopher Columbus defined bureaucracy while discovering America: he did not know where he was going, he was not sure how he was going to get there, and he did not recognize it once he arrived. Unfortunately, the same can often be said for fire service organizations—especially when they are called upon to face the ever-changing issues and the demand for change common to all modern organizations.

Dealing with the issues and changes involved in developing and maintaining a quality management program in the fire service involves a systematic process that has four stages: analysis, planning, implementation, and evaluation. Initial analysis involves examining the existing situation to determine the changes that need to be made. Methods and activities useful for analysis were discussed in Chapter 2: Overview of Quality Management.

This chapter will focus on the second Baldrige category: strategic planning. This involves using information gained during the initial analysis to develop a plan that addresses the needs identified.

◆ STRATEGIC QUALITY PLANNING

Strategic planning is the process of developing both long- and short-term organizational objectives, ways to achieve those objectives, and methods to measure their effectiveness. *Quality planning* is the process of designing the system to perform to the standards expected by customers or other stakeholders. Strategic and quality planning can and should be combined into a single agency-wide planning process (Campbell, 1993). Simply put, strategic quality planning is an organized method of determining where an organization wants to be and how it plans to get there. It should be an integral and ongoing part of all the operations of the entire organization.

THE TEAM APPROACH

Teams routinely handle day-to-day operational issues in a wide range of business arenas. Businesses use teams to solve customer-service problems, identify purchasing options for equipment, research technological advances, and determine better methods of providing service.

Teams are effective at accomplishing things because of the combined expertise found in a group. No one person has all the answers, but when the knowledge and skills of several people are brought together, the results can be powerful. Improved decision making, faster problem solving, and greater productivity are a few of the potential outcomes.

Rarely does the fire organization provide services in a void. In other words, regardless of the service provided, other organizations and stakeholders are involved and affected. The fire service cannot operate without dispatchers. Major fires, hazardous materials incidents, and other crises always involve a contingent from law enforcement and may include representatives from other groups such as OSHA. Rescue may involve organizations such as the local power company, a local military contingent, or OSHA. EMS involves interacting with hospitals. Using integrated teams with representatives (both administrative and line) from organizations that are affected by fire operations helps identify potential solutions that address concerns of all entities involved. Increasing member involvement through the use of teams also leads to improved job satisfaction. When members develop a sense of ownership over their decisions, they become committed to carrying them out.

The Park City, Utah Fire District used an Incident Management Team (IMT) to manage planning and operations during the 2002 Winter Olympics. IMTs are used across the country to manage all types of major incidents; the team concept can be used to plan for and manage large or complex incidents or major changes within the department. *(Photo by Gordon Sachs.)*

Teams are not a method to "appease the masses." Nothing is more frustrating than to spend hours and days developing a plan, only to have it thrown out without consideration because a decision has already been made at the upper level. Teams need a solid commitment from leadership and management. The leadership of the organization must trust and respect the teams' abilities and allow them to carry out their objectives without interference. Teams must be provided with the resources they need to get the job done. These resources include a place to meet, funds to support the teams' work, equipment and supplies, access to data and information, and time to do the job.

◆ IDENTIFYING THE TEAM

Before determining who will serve on a team, the leader must first identify what the team is to do. For example, developing the organization's vision and mission statement is directed by the chief officer and may involve mostly upper management, with one or two representatives from middle management and line personnel. Developing objectives and performance standards should involve more line personnel because they actually do the job, know what it entails, and know the potential problems that might be encountered. All line personnel chosen to serve on a team should have the respect of their peers. This is a tremendous asset in gaining buy-in from the entire workforce. For teams to be effective, there should be a minimum of 5 members and a maximum of 12 (United States Fire Administration, 1999).

Someone must be identified to facilitate the team. Often, the initial facilitator is a middle manager. This person must facilitate the team effort by keeping members on task, not directing them to preconceived conclusions. Ideally, the leadership role should rotate among members—regardless of rank—as the team develops.

Before the team begins its work, the leader should hold an initial meeting to explain the purpose of the team and the goals to be accomplished. Team meetings should be frequent enough (once every week or two weeks) to allow progress to occur. All team meetings should be held on duty. Teams may become a permanent part of the management system, such as the Safety and Health committee, an EMS QI team, and a disaster planning team. Teams may also be formed for a specific purpose, such as an apparatus specification committee or protective clothing committee, and exist only until a project's conclusion.

◆ STEPS IN THE PLANNING PROCESS

Strategic quality planning is an ongoing process, based on the principles of quality improvement discussed in Chapter 2. Developing a complete strategic quality plan involves the organizational, financial, and functional aspects of the service. This planning is not a separate process from other organizational planning. Therefore, it must be integrated with existing planning and quality methods such as OSHA regulations and National Fire Protection Association (NFPA) standards. Adding new or revising existing services based on customer needs and expectations should follow a specific order of steps (Figure 4.1).

Developing the vision statement is the first step. The vision statement was discussed in Chapter 3. However, it bears repeating that the vision statement serves as the foundation for the rest of the planning process. Every objective and action plan identified in the planning process must be consistent with the vision statement.

Figure 4.1 ◆ Strategic Quality Planning Steps.

- Develop a vision statement.
- Identify underlying assumptions.
- Identify key drivers.
- Develop objectives and performance indicators.
- Determine compliance for performance indicators.
- Develop and implement action plans.
- Evaluate effect of action plans on performance indicators.
- Modify action plans or performance indicators as needed.

Underlying any planning process are implicit **assumptions** that guide the organization in certain directions. Identifying and discussing these assumptions is important to determining whether they are still valid or whether they need to be changed. The following assumptions serve as examples:

- The public expects the fire department to handle their emergencies.

This may seem like an obvious assumption until it is analyzed. The key to this assumption is identifying the difference between how the fire organization defines emergency and how the public defines emergency. How many pictures have you seen of a firefighter on a ladder rescuing a cat from a tree? The organization no doubt does not define a firefighter's rescuing a cat from a tree as an emergency, but to the little old lady whose cat is stuck, it is an emergency. Discussing this assumption may result in changing the organization's policy on what they respond to or in the organization's developing a specific public education program.

- Changing technology and community needs significantly impact the evolution of services provided by a fire organization.

This assumption has profound effects on several areas within the organization, including equipment, training, and budget. Does the department have a regular equipment replacement schedule? Is there a capital equipment fund that carries over from year to year to allow regular replacement of expensive apparatus without the need for a voter-approved bond issue? What new services should be considered over the next five to ten years? What will those new services require in terms of personnel, equipment, and training? What is the projected cost?

Key drivers are those performance areas most critical to the success of the fire organization. They should be consistent with the vision and mission statement. The key drivers are the basis for quality improvement efforts in specific areas.

Identifying key drivers is based on expert opinion, good judgment, and common sense. Because fire organizations provide a multitude of services in many areas, it is important to involve representatives of each division in the process of identifying key drivers. After the team identifies the drivers, they should be validated by both internal customers (employees and volunteers) and external customers (public focus group, citizens committee, and other organizations such as law enforcement, dispatchers, etc.). Figure 4.2 provides some examples of key performance areas for a fire organization. This list is meant as an example and is not comprehensive.

- Prevention
- Appropriate scene response times
- Effective fire suppression
- Effective management of hazardous materials incidents (Hazmat Division)
- Timely and appropriate patient intervention (EMS)
- Customer satisfaction
- Workforce health and safety
- Workforce relations

FIGURE 4.2 ◆ Fire Service Key Drivers.

◆ OBJECTIVES AND PERFORMANCE INDICATORS

Objectives are measurable statements that are consistent with the organization's mission, vision, and key drivers. There are two types of objectives. **Procedure-oriented** objectives are those that are assumed to help achieve the broad plan objectives. **Outcome-oriented** objectives provide the means to measure accomplishment. Figure 4.3 provides an example of a procedure-oriented objective with a related outcome-oriented objective.

The objectives defined in the strategic planning process often start out as broad organizational goals. To be effective, the objectives must be clearly defined. If they are well-defined, some objectives can also be used as performance indicators. However, often these objectives are too broad to use as performance indicators. For example, a fire organization has the objective of providing the safest possible environment for personnel operating on the fire ground at all times. This is a very broad objective. Additional indicators must be identified to measure the attainment of this objective and analyze any problems with compliance.

Performance indicators are the actions that measure the attainment of the objective. They must be specific, attainable, and realistic and should be based on governmental regulations, such as OSHA, and national standards, such as NFPA 1500. Using the objective above as an example, some performance indicators might include the following.

- The Incident Command System is followed.
- The two-in, two-out requirement is followed in all Imminently Dangerous to Life and Health (IDLH) situations.
- Rehabilitation procedures follow department SOP.
- All department apparatus meet appropriate standards set by the NFPA.
- Apparatus is maintained through a regular preventive maintenance program described by the manufacturer.
- All personal protective equipment meets appropriate NFPA standards.

Procedure-oriented: Finalize computerized response time data collection and reporting system by January 2002.

Outcome-oriented: Fire apparatus will arrive at the scene of a call within 5 minutes 90% of the time.

FIGURE 4.3 ◆ Objectives.

In order to measure how well the organization is meeting this objective, performance indicators must be written with a **compliance level** (expected outcome) stated. The following examples include compliance levels.

- The Incident Command System is followed *at all incidents.*
- The two-in, two-out requirement is followed *in all IDLH situations.*
- Rehabilitation procedures follow department SOP *90%* of the time.
- All department apparatus meet standards set by the NFPA *100%* of the time.

Because the analysis of compliance levels is used to determine the extent to which objectives have been met, consideration and forethought must be given to setting those levels. If all compliance levels are set at 100% effectiveness, any program is doomed to failure. It is easy to think that the compliance level for the indicators used as examples should all be set at 100%. However, extenuating circumstances such as multiple incidents that utilize all available EMS personnel may prevent standard rehabilitation procedures from being met 100% of the time.

Once the indicators are written, they should be reviewed with a critical eye. Our example indicator that "All department apparatus will meet the standard set by NFPA 100% of the time" poses two problems as written. The first is the indicator itself. Specifying "all" does not allow for any apparatus to be out of service or down for maintenance. Thus, the indicator would be better written "All department apparatus *in service* meet standards. . . ." In addition, the compliance level should not be set at 100%: equipment required as part of the standards may be out of service while the apparatus is still in service.

There are two primary methods for determining appropriate compliance levels. The first is benchmarking. **Benchmarking** uses comparisons based on data from other organizations, states, or regions. Benchmarking may also use national standards such as those set by OSHA regulations, NFPA standards, and state law. Using personal protective equipment and following universal precautions when there is potential exposure to any body fluid 100% of the time is a national standard. This is an objective of the organization's infection control program. Again, this is broad, so performance indicators should be developed that identify situations that require just gloves and those situations that require additional protection such as personal respiratory protection, eyewear, and so forth. If there is no standard or accepted compliance level for an objective/performance indicator, the organization must determine the compliance level through data collection and analysis.

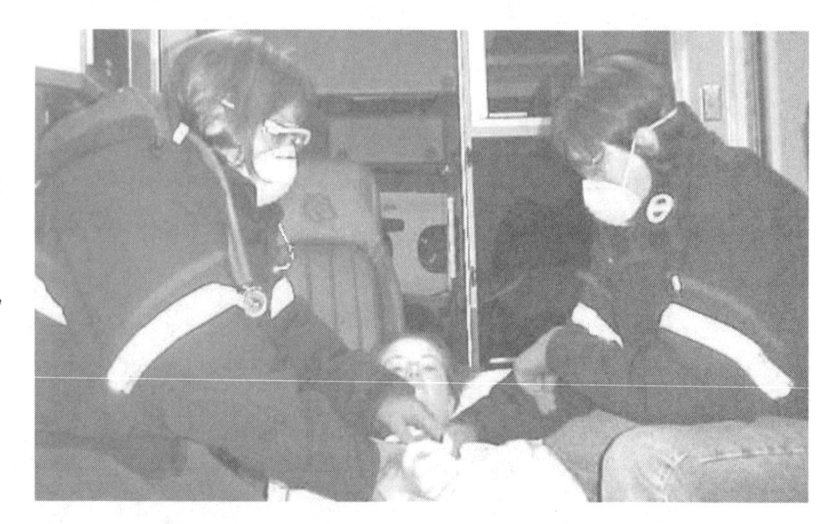

Using personal protective equipment and following body substance isolation practices when there is potential exposure to body fluids 100% of the time is a national standard and can be a benchmark for a department's infection control program. *(Photo by Judy Janing.)*

◆ ANALYSIS

When compliance levels are not met, an analysis needs to be completed to determine the cause of the problem. A fire organization is a highly complex system. The analysis, as well as the action plans that result, must take this complexity into account. Analyzing the cause of noncompliance will likely involve the entire organization, including policy makers, line personnel, health and safety officer, infection control officer, risk manager, and others.

Let us refer back to one of the example performance indicators used earlier. The fire organization has established that the two-in, two-out requirement will be used for all IDLH atmospheres (100% compliance). However, compliance is determined to be only 60% of the time. The analysis of the situation may identify any one or a combination of the following reasons for why not enough firefighters arrive on the fireground in a timely manner (the non-compliance.)

- Fire stations are located too far apart.
- An inadequate number of units were initially dispatched on the response.
- Some responding personnel were inadequately trained.
- Retirements/sick times have caused a shortage of personnel.
- The city has implemented a downsizing policy or hiring freeze.
- The city administrator/city council has put political pressure on the fire chief to reduce overtime.
- There is inadequate funding in the budget to pay for overtime needed to meet the minimum staffing standard.
- The fire chief does not believe that minimum staffing is really necessary.
- Personnel are assigned at the beginning of a shift to meet the minimum staffing requirement, but are then detailed to other divisions.
- Members of the volunteer department are not available in the daytime during working hours.

Each reason may ultimately require action steps at different levels and may involve more than one division of the organization.

◆ STRATEGIC FINANCIAL PLAN

Once the organization has developed their vision and mission statements and identified needed services, the inevitable question emerges: What are these services going to cost? The strategic quality plan must be accompanied by a strategic financial plan. Fire organizations can no longer afford to operate on just a yearly budget.

For example, many career and volunteer fire departments count on reimbursement from their EMS component to offset expenses. With the advent of managed care, the fee-for-service form of reimbursement is being revised to one of capitated payment. In addition, the Centers for Medicare and Medicaid Services (CMS) rules regarding ambulance reimbursement for Medicare patients have changed dramatically.

The fire organization can no longer decide which services it wants to deliver without assessing the economics of demand. Strategic financial planning must go hand in hand with the strategic quality planning. If the strategic quality plan calls for the addition of new services and those services cannot be supported in the strategic financial plan, the organization may have to review and redefine its mission statement and revise the overall strategic plan.

◆ REQUIREMENTS FOR EFFECTIVE FINANCIAL PLANNING

Figure 4.4 summarizes the requirements for an effective strategic financial plan. These include the following.

- Data on cost, revenue, and investment should be available along program lines.

Unfortunately, many fire organizations' accounting systems are geared to provide data along departmental lines. Municipal system data is often part of the overall city budget, and within the organization, data is often confined to overall costs versus reimbursement. The organization should be able to break out revenue and cost by both division and by each program that the division oversees.

- Working capital is a major element.

New programs may have significantly different working capital requirements. For instance, providing a Hazardous Materials unit for use by surrounding communities requires significant fixed investment in terms of building space, equipment, and training. This requires a considerable amount of working capital to finance a future revenue source and initial development costs. If inadequate amounts of working capital are projected, the entire financial plan may be in jeopardy.

- There should be some accumulation of funds for future investments.

Set-aside funds for investment to meet future needs should be approached in the same manner that pension plans are funded.

- Establish a limit on debt financing.

The level of indebtedness required to finance the strategic plan might expose the organization to excessive risks. If those risks are not recognized until actual financing is needed, existing programs may have to be cut back or cancelled to free up funds for more desirable programs.

- Return on investment is an important factor in selecting programs.

Determining return on investment revolves around categorizing programs in terms of community need. Return on investment is the measure of profitability. However, it cannot be viewed only in terms of revenue. It must also be viewed in relationship to the underlying community need.

FIGURE 4.4 ◆ Strategic Financial Plan Requirements.

- Track data on cost, revenue, and investment program lines.
- Identify source of working capital.
- Accumulate funds for future investments.
- Establish a limit on debt financing.
- Determine return on investment when selecting programs.
- Include non-operating sources of equity.
- Review and update the plan yearly.

Community need may be difficult to quantify. Organizations need to look at both existing services and proposed new programs to determine if they are duplications of services provided by other organizations or if they can augment and improve existing services. This may involve partnering with various outside groups such as vendors for smoke detector programs, health agencies for immunization programs, or safety councils for car seat programs.

Programs should be reviewed to determine if they are high or low community need. Obviously, the organization wants to find programs that have a high community need, thus a high return on investment. However, if community need is high, certain programs may be implemented even though the return on investment is low. An example of this type of program might be the smoke detector program. This program is normally offered free of charge. Although a vendor or other organization may provide the detectors for free, the personnel and wages expended to install them in homes are an expense that is not recovered. Programs that have low community need and low return on investments should be reviewed to determine if they will be continued.

- Include non-operating sources of equity.

The primary source of non-operating income is investment income and gains. These investments may include retirement plans, professional liability plans, funded depreciation, bond funds, or endowments.

- The plan should be reviewed and updated yearly.

The organization does not expect responding apparatus to find addresses based on five-year-old maps and information. Similarly, the organization should not expect to operate effectively and efficiently with an outdated financial plan.

◆ ACTION PLANS

DEVELOPING THE PLAN

Action plans are necessary to bring new services or programs on-line and to correct problems identified when the organization is out of compliance with performance indicators. Action plans should be specific, identifying the exact steps necessary to correct the problem, the time schedule to accomplish the plan, and the methods used to evaluate the effectiveness of the plan. In addition, obstacles that might impact the implementation of the plan and resources needed to carry out the plan need to be identified.

An effective action plan is dependent on the team that develops it. In addition to the appropriate management representative, team composition should include representatives directly affected by the performance indicator, including line personnel. A time frame for both the completion of the action plan and the accomplishment of the actions should be identified. Providing the team with a schedule and checkpoints for completion of the plan helps keep them on-task to reach a solution in an acceptable time frame.

IMPLEMENTING THE PLAN

In addition to the obvious necessity of allocating resources, the biggest challenges during the implementation phase are gaining buy-in from personnel and maintaining focus and momentum among team members. Gaining buy-in often hinges on the prevailing organizational culture. It is important to consider the culture because it reflects the basic values of the organization.

The fire service holds some undeniable values, the foremost being to save lives and property. This is the tenet on which the fire service was founded. As the fire service has matured, other values have emerged and are emerging. These include critical thinking and participatory management. These two values may lead to conflict and frustration. Thus, although these are values of today's fire service members, they may not be shared by a fire service leadership that continues to function in a very centralized management structure. This is true in both career and volunteer fire departments.

Shared values are the dominant beliefs about what is important that underlie the organization. These may be values related to a variety of areas:

(a) key competency of the organization, such as providing comprehensive public education programs;

(b) operational performance, such as response times; and

(c) organizational output, such as skills proficiency.

Norms are the informal rules that influence decisions and actions throughout the organization. Norms reflect what is accepted as appropriate and inappropriate behavior. Norms should encourage behavior consistent with shared values and organizational mission. Norms are strongly tied to the activities of the organization's leaders—how well they "walk the talk." Regardless of the stated values, the norms are based on demonstrated behaviors.

Motivation is also key to gaining acceptance of change by members. It is human nature to resist change, even if the change is good. Change requires adaptation and new learning and takes everyone out of his or her comfort zone. Making line personnel part of the team during the development and implementation stages and linking the change to the organization's mission and objectives can go a long way in furthering the acceptance of change.

Maintaining focus and momentum among team members requires three simple actions.

1. **Follow the schedule.** This may seem rather obvious, but often is not done. Using the schedule helps concentrate efforts and attention where they are most needed. It also helps anticipate the next steps. Schedules aid coordination and communication within a team. Everyone should be able to identify where he or she is at any given point.

2. **Keep others informed.** This would also seem obvious, but is often neglected or ignored during critical periods. Schedule communication opportunities, and carry through.

3. **Empower the team.** Provide the resources and authority to accomplish the task, and *allow* the team members to do what you have charged them to do.

EVALUATING THE PLAN

After the plan is implemented, it is critical that the results of the implementation be monitored and evaluated to determine results or problems that arise. The evaluation phase focuses on measuring the difference between what was happening before and what is happening now. Therefore, the evaluation of an action plan implemented to correct noncompliance with a performance indicator requires that compliance be reevaluated.

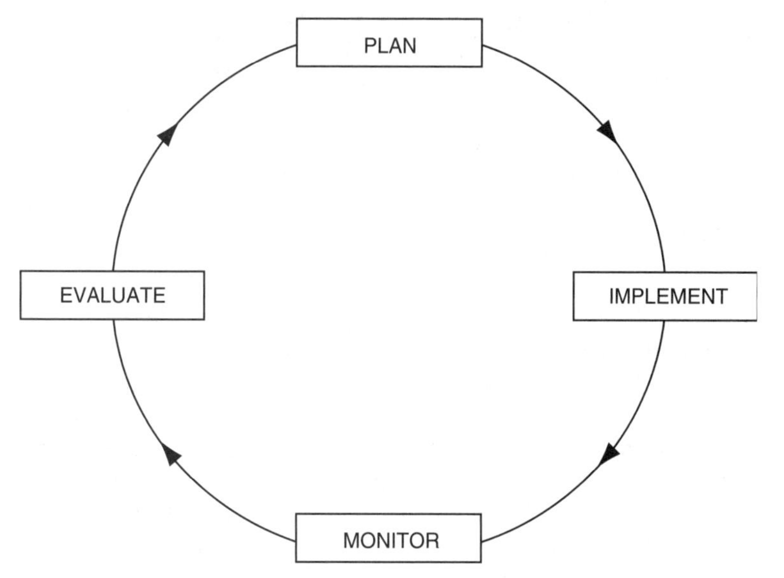

FIGURE 4.5 ◆ Formal Evaluation Process.

Any program is incomplete without a formal evaluation process. A formal evaluation process is cyclical (Figure 4.5). Although evaluation is the "final" step, it actually leads to a repetition of the cycle.

Monitoring and evaluation may take a variety of forms depending on the actions specified in the plan. Methods to collect data to monitor results should be developed during the planning stage. Chapter 5 covers relevant data elements in depth. In addition to data collection internally, the organization may use the same sources of data discussed in Chapter 3: Customer Focus Groups and Surveys.

It is important that the monitoring methods chosen allow for timely feedback so any problems will be detected early. If problems with the plan are detected during the monitoring phase or the action plan did not make a difference, the organization must revise. Based on the problem identified, revision may require re-entering the process cycle at any point.

CHAPTER SUMMARY

Strategic quality planning is an organized method of determining where an organization wants to be and how it plans to get there. It is a systematic process that has four stages: analysis, planning, implementation, and evaluation. It entails identifying the organization's key drivers, developing objectives and performance standards, determining and monitoring compliance levels, and developing, implementing, and evaluating action plans. Strategic financial planning must be considered in conjunction with the strategic quality plan to assure needed resources.

Successful implementation of the strategic quality plan requires buy-in from the workforce. "Walking the talk," relating the plan to the vision and mission statements, and using the team approach in the development of both the strategic quality plan and action plan promote ownership and acceptance of change.

Strategic Planning Checklist

[ ] **Use the team approach.**
- [ ] Identify what the team is to do.
  - Will this be a permanent team or one formed for a specific purpose?
- [ ] Identify the team facilitator.
- [ ] Meet with the team and explain its purpose and goals.
- [ ] Assure regular team meetings (once every week or two weeks).

[ ] **Follow the steps in the planning process.**
- [ ] Assure that the vision statement is current.
- [ ] Identify the underlying assumptions.
  - What does the public expect of the organization?
  - What are the effects of changing technology?
  - What are the effects of changing community development and needs?
- [ ] Identify the key drivers.
  - What performance areas are most critical to the success of the organization? (See Figure 4.2.)
- [ ] Validate the drivers with internal and external customers.
- [ ] Write objectives and performance indicators.
- [ ] Determine compliance level.
  - Use benchmarking or data analysis.
- [ ] Determine evaluation method.
  - Determine data to be collected and methods of collection.
  - Determine review time frame.
  - Determine method of analysis and who will analyze.

[ ] **Complete the strategic financial plan.**
- [ ] Obtain data on cost, revenue, and investment along program lines.
- [ ] Determine working capital.
- [ ] Identify investment funds.
- [ ] Establish limit on debt financing.
- [ ] Determine return on investment.
  - Is the community need for this program high or low?
  - Is the program proposed a duplication of services provided by other agencies?
  - Is there an agency that the organization can partner with to provide the service?
- [ ] Identify non-operating sources of equity (e.g., retirement plans, bond funds, etc.).
- [ ] Review and update plan yearly.

[ ] **Develop the action plan.**
- [ ] Select appropriate team members.
- [ ] Determine the time frame for completion of the action plan and the accomplishment of the actions.
- [ ] Develop the schedule and checkpoints for plan completion.

[ ] **Implement the action plan.**
- [ ] Allocate resources.
- [ ] Gain member buy-in through effective communication and member input.
- [ ] Follow the schedule.
- [ ] Keep others informed.
- [ ] Provide the team with the resources and authority to accomplish the task.

[ ] **Evaluate the plan.**
- [ ] Review and analyze data on results to allow for timely feedback.
- [ ] Revise the plan based on the problem(s) identified.

Activities

1. Identify key drivers for your organization. Relate the key drivers you identify to your vision and mission statements.
2. Develop objectives or performance indicators for at least three of the key drivers you identified in Activity 1. Include compliance levels. If you use a national standard, reference the standard used.
3. Identify the similarities and differences between your organization's financial planning and the strategic financial planning process discussed in this chapter.
4. For your organization, discuss the following.
   a. What are the shared values and how are they communicated?
   b. What are the norms?
   c. What is the dominant management style?
   d. Do the norms reflect the shared values? Explain.
   e. Does the dominant management style support the shared values? Explain.

References

Campbell, A. B. (1993). Strategic planning in health care: methods and application. *Quality Management in Health Care* 1(4), 12–23.

United States Fire Administration, National Fire Academy. (1999). *Advanced leadership issues in EMS.* Washington, DC: U. S. Government Printing Office.

Bibliography

Aaker D. A., & Adler, D. A. (1998). *Developing business strategies* (5th ed.). Indianapolis, IN: John Wiley & Sons.

Cleverley, W. (1997). *Essentials of health care finance* (4th ed.). New York, NY: Aspen Publishers.

National Highway and Safety Administration. (1997). *A leadership guide to quality improvement for emergency medical services.* Washington, DC: U.S. Government Printing Office.

Human Resource Development and Management

5 CHAPTER

◆ INTRODUCTION

It would be extremely difficult, if not impossible, to find a chief officer who would not identify personnel as the most important asset of the fire service organization. The majority of fire service members want to do a good job. Yet there are endless stories of unsatisfied or "problem" workers. If the organization policies and general orders were reviewed, many could be attached to a specific name. Too often, the energy, effort, and time of the organization's leaders are focused on directing and managing that small percentage of workers who are perceived as problems.

This current state of affairs may be due, in part, to the quasi-military structure of the fire service. This traditional structure has carried over and reinforced the discipline-centered and "follow-orders" approach used on the fire ground in day-to-day operations; in addition, it has tempered the creativity of line personnel in areas of organizational improvement.

There is no question that the fire ground requires a structured incident management system that follows the military command and control structure, but the routine, day-to-day management of a fire organization should foster and support line personnel involvement in improving services by being open to new ideas and approaches for meeting the needs of the customers served. Neither the noblest vision of a chief officer nor the best intentioned organizational mission statement will be fulfilled if the workforce does not share the vision and is not empowered to develop and use their full potential to achieve goals.

To use the workforce to their full potential, the chief officer and upper management must begin by evaluating the current human resource conditions in relation to the organization's strategic plan. The results of this evaluation serve as the basis for planning changes where needed. Human resource evaluation involves reviewing the current (a) workforce work systems, (b) education, training, and personal development, (c) safety and health, and (d) well-being and satisfaction.

HUMAN RESOURCE EVALUATION AND PLANNING

Human resource evaluation focuses on assessing and improving human resource planning, practice, and performance. The fire organization's leaders must carry the goals identified during the strategic planning process into the realm of human resources. This involves linking specific quality goals to specific human resource goals and identifying what human resources must be in place to help ensure the success of the strategic goals. For example, a goal to implement a hazardous material response unit must consider existing skills, capabilities, and knowledge of the workforce as well as the need for new equipment and technology. Aspects that must be reviewed for potential change include the following.

- SOPs related to efficiency, coordination, and response time intervals
- workforce development, education, and initial and refresher training (including credentialing)
- compensation, recognition, and benefits
- staff composition

Determining the current state of human resources practices involves gathering both personnel-related and organizational performance data and analyzing that data for links between human resource practices and key performance results. The analysis should then be used to determine changes needed to achieve the system goals identified in the strategic plan. Types of data to analyze include member job satisfaction surveys, turnover, absenteeism, safety records, grievances, recognition programs, training records, and exit interviews. Overall system strengths and weaknesses, such as city or county personnel policies, that could affect the organization's ability to fulfill a proposed human resource plan must also be reviewed. This may require involving the personnel or legal departments of the city or county.

After the initial evaluation is completed, the human resource plan should be developed in the context of the strategic plan. Human resource planning should focus on the following initiatives.

- redesigning the work process or jobs to increase opportunity, responsibility, and decision making for line personnel
- promoting labor-management cooperation
- recognizing and rewarding efforts that increase customer satisfaction
- soliciting input from personnel to identify ways to improve performance
- prioritizing personnel problems based on potential impact on productivity
- developing recruitment and retraining strategies
- forming partnerships to increase education, training, and job opportunities
- addressing safety factors

After implementing any changes, it is critical to evaluate the effect of those changes on performance and satisfaction. In addition to collecting and comparing the results of the internal data sources identified above, fire organizations can also compare their evaluations to those of organizations providing similar services and use pre-established benchmarks to identify specific personnel needs or new approaches or practices.

◆ WORKFORCE WORK SYSTEMS

Improving the quality of services offered by the organization may require that the workforce be reorganized into different, more effective work units. This may mean developing problem-solving teams or formal and informal functional units that may be temporary or long-term. These work units may cut across customary organizational lines and be either self-managed or managed by supervisors.

Job performance can be improved when each job description ensures that roles, responsibilities, duties, and tasks are tailored to achieve the organization's goals. This may involve simplifying job classifications, cross-training, rotating jobs, modifying work locations, or using new technologies such as computer links or conferencing.

Leadership must create opportunities for initiative and self-directed responsibility for line personnel involved in various work units. Initiative and self-direction calls for the organization's leaders to foster flexibility, job efficiency, task coordination, and the ability for the work team to respond quickly to changing requirements. This requires the freedom to effectively communicate across traditional units or divisions. For this to be successful, leadership may need to work with the local labor unit to obtain buy-in and support, and may require changes in the current organizational culture.

This is no easy task. The "release" of power by chief officers is critical to the success of improving workforce systems. This requires a great deal of confidence in both their abilities and the abilities of the members. The fire service is steeped in paramilitary tradition and it is difficult to delegate the authority that comes with rank. Many paid departments function on the seniority system for assignments. Many volunteer departments use the election system for leadership positions; anticipation of future successful elections can have a profound effect on whether or not authority is delegated. As with any change, the success of this endeavor begins with the commitment of the chief and upper management and should occur over time.

Changing the organizational culture begins with one small step. Using the team approach as discussed throughout this book and truly empowering that team to make decisions is a start. Chief officers must be willing to allow the team to implement their recommendations and even fail, if that is the result. Failure in noncritical areas can be one of the most beneficial learning experiences available. This allows the team to be accountable for their decisions, generate new solutions, and most importantly, reinforces management's commitment to workforce input.

Success in small projects builds trust between management and members. Management learns that delegating authority not only lessens their duties, but also reaps new and innovative approaches. Members learn that management really is willing to let them contribute to the improvement of the organization. Moving into areas such as cross-training and reorganization requires trust and a strong working relationship with members and member organizations.

◆ WORKFORCE EDUCATION, TRAINING, AND DEVELOPMENT

Education/training is a primary tool for empowering the workforce to both meet job requirements and achieve the mission and vision of the organization. The majority of training that occurs within the fire organization is directed at meeting practice skills, whether they are fire suppression skills or EMS clinical skills, and certification requirements. However, in the context of the strategic plan, education and training should extend beyond the need for practice expertise. For example, in addition to possessing content mastery, training instructors should be knowledgeable in both adult learning characteristics and teaching methodology. Supervisors and team members should be trained in the skills relevant to team activities such as leadership and team facilitation. The chief officer might consider working with state agencies or local community colleges and universities to organize and develop curriculums specific to the needs of the fire organization and to arrange multijurisdictional classes.

Fire personnel interact directly with the organization's external customers on a daily basis. Training for customer contact should include areas that contribute to customer awareness, such as effective listening skills, how to gather relevant information from customers, managing customer expectations, and anticipating and handling system problems or failures. Personnel should also be educated on the importance of treating all people with respect.

Analyzing job responsibilities and the types and levels of skills required can be done to determine these nontraditional training needs. Improving the education, training, and development of personnel should also include involving the workforce in determining specific education and training needs, evaluating and selecting appropriate delivery and evaluation options, and determining how knowledge and skills are reinforced on the job.

Alternatives to the standard classroom setting delivery method should also be considered. Many fire organizations have access to a dedicated television channel. Training programs (including interactive programs) are available on CD-ROM, and distance education via the Internet is increasing on a daily basis. These formats allow for maximum scheduling flexibility for a workforce whose duty responses are unpredictable.

Another source of educational courses is the United States Fire Administration, National Fire Academy (NFA). Although the number of students able to attend on-campus courses is limited, NFA is beginning the process of releasing residential, regional, and volunteer incentive program courses to states for delivery. The concept is called *endorsement.* It recognizes that state training systems are extensions of the NFA in their state and the courses result in the issuance of an NFA certificate.

The U.S. Fire Administration's National Fire Academy offers training in a variety of formats, including computer-based training on CD-ROMs. These are available to fire departments at no cost. *(Photo by Gordon Sachs.)*

States must meet two requirements in order to deliver these courses.

1. The courses must be taught only by NFA instructors from their qualified bid list.
2. NFA instructors are required to teach the same number of hours as the residential (on-campus) course.

NFA resident courses are accredited by the American Council on Education (ACE) for three upper division undergraduate college credit hours.

◆ WORKFORCE HEALTH AND SAFETY

The health and safety of the workforce is a primary concern of several governmental agencies as well as fire service organizations.

OSHA

The Occupational Safety and Health Administration (OSHA) publishes federal regulations that establish minimum standards for workplace safety and health. Emergency service personnel are not always covered by OSHA regulations. Determining whether or not a particular agency is required to comply with OSHA regulations requires research.

In general, federal OSHA regulations apply only to federal, military, and private employers. Thus, in states that have simply adopted the federal OSHA regulations, most emergency service personnel (whether paid or volunteer) are not covered.

If a state opts to develop its own OSHA plan, all paid state and local government members (including emergency response personnel) must be covered. (In some states, only state and local employees are covered.) However, each specific state decides whether or not to include volunteer emergency response personnel. Figure 5.1 shows the states that currently have state OSHA plans.

Emergency response personnel involved in hazardous material response in states and territories having a state OSHA plan are covered under OSHA 29 CFR Part 1910.120. For workers not covered by a state OSHA plan, the Environmental Protection Agency (EPA) issued a regulation (40 CFR Part 311) in 1989. This regulation is functionally similar to OSHA's Hazardous Waste Operations and Emergency Response

Alaska	Michigan	South Carolina
Arizona	Minnesota	Tennessee
California	Nevada	Utah
Connecticut	New Jersey	Vermont
Hawaii	New Mexico	Virginia
Indiana	New York	Virgin Islands
Iowa	North Carolina	Washington
Kentucky	Oregon	Wyoming
Maryland	Puerto Rico	

FIGURE 5.1 ◆ States/Territories with OSHA Plans.

regulation, specifically citing that the requirements in the OSHA 29 CFR Part 1910.120 will be applicable in all states, regardless of OSHA status. Communicable disease risks and hazards are the same in each state, regardless of local compliance requirements. Thus, adherence to OSHA regulations is recommended for all emergency service agencies.

Fire organizations must consider several regulations and standards when developing their quality management plan. For example, *Respiratory Protection* (29 CFR Part 1910.0134) outlines the two-in, two-out requirement. Essentially, this requires that two trained and equipped members of a fire department who are capable of making an emergency rescue of other firefighters be outside a hazardous area anytime there are two (or more) firefighters operating in the hazardous environment. OSHA defines the hazardous environment as any area where the atmosphere is "imminently dangerous to life or health" (IDLH), such as inside a burning building. The only exception to this requirement is when there is an obvious rescue situation where immediate action will result in saving a life.

The two-in, two-out rule is based on OSHA's Respiratory Protection regulation and is addressed in NFPA standards. The intent is to reduce the risk to firefighters during the initial stages of an incident. *(Photo courtesy Gordon Sachs.)*

The Occupational Exposure to Bloodborne Pathogens (29 CFR Part 1910.1030) regulation establishes standards for workplace protection from bloodborne diseases. The primary diseases of concern are hepatitis B (HBV) and Human Immunodeficiency Virus (HIV). The primary methods of protection are training, engineering and work practice controls, immunization against HBV, and the use of body substance isolation.

This standard has been updated as a result of the Health Care Worker Needlestick Prevention Act (PL 106-430). Occupational Exposure to Bloodborne Pathogens, Needlesticks, and Other Sharps Injuries: Final Rule (29 CFR Part 1910) requires that employers utilize needleless systems and sharps with engineered injury protections to prevent the spread of bloodborne pathogens. It also requires changes to the exposure control plan, requires the maintenance of a sharps injury log, and addresses training changes.

The Centers for Disease Control (CDC) publishes the *Morbidity and Mortality Weekly Report* (MMWR), a weekly update of information on communicable diseases. The infection control officer should periodically review MMWR for new information. Some of the recommendations that specifically address public service personnel (including firefighters and EMS) include the following.

- *Recommendations for Prevention of HIV Transmission in Health-Care Settings* (MMWR Vol. 36, No. 2, August 21, 1987) This document established the concept of "universal precautions." Under universal precautions, the blood and certain body fluids of *all* patients are considered potentially infectious.
- *Guidelines for Prevention of Transmission of Human Immunodeficiency Virus and Hepatitis B Virus to Health Care and Public Safety Workers* (MMWR Vol. 38, No. 2-6, 1989) This

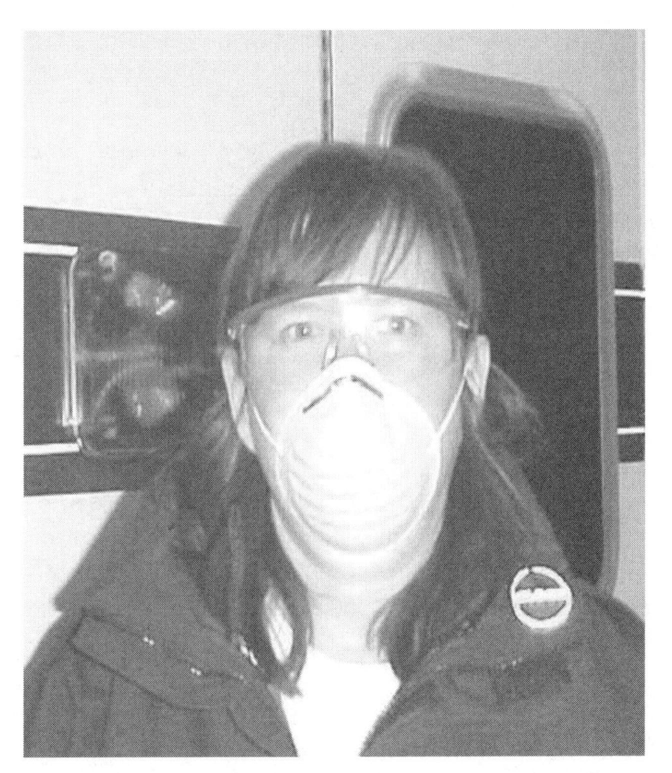

Protection from bloodborne pathogens is an important part of a fire department's health and safety program, and is addressed in OSHA regulations and NFPA standards. *(Photo by Judy Janing.)*

document expanded universal precautions to the extent that "under uncontrolled, emergency circumstances in which differentiation between fluid types is difficult . . ." all body fluids should be considered potentially hazardous.

- *Recommendations for Preventing Transmission of Human Immunodeficiency Virus and Hepatitis B Virus to Patients During Exposure-Prone Invasive Procedures* (MMWR Vol. 40, No. RR-08, July 12, 1991) This document addresses the status and outlines the scope of practice of health care workers who have HIV or HBV.
- *Guidelines for Preventing the Transmission of Mycobacterium tuberculosis in Health Care Facilities, 1994* (MMWR Vol. 43, No. RR-13, October 28, 1994) This document describes the requirements for personal respiratory equipment and guidelines specific to EMS regarding transport of potential TB patients.
- *Immunization of Health Care Workers: Recommendations of the Advisory Committee on Immunization Practices (ACIP) and the Hospital Infection Control Practices Advisory Committee (HICPAC)* (MMWR Vol. 46, No. RR-18, December 26, 1997) This report discusses all the immunizations recommended for health care workers and provides the regime for those immunizations. ACIP strongly recommends that all health care workers be vaccinated against (or have documented immunity to) hepatitis B, influenza, measles, mumps, rubella, and varicella (chickenpox).
- *Public Health Service Guidelines for the Management of Health Care Worker Exposures to HIV and Recommendations for Postexposure Prophylaxis* (MMWR Vol. 47, No. RR-7, May 15, 1998) This report describes recommendations for the management of those who have occupational exposure to blood and other body fluids that may contain human immunodeficiency virus (HIV).
- *Recommendations for Prevention and Control of Hepatitis C Virus (HCV) Infection and HCV-Related Chronic Disease* (MMWR Vol. 47, No. 19, October 16, 1998) This document addresses management guidelines for Hepatitis C (HCV) and recommends that emergency medical, and public safety worker institutions establish policies and procedures for HCV testing of persons after exposures to blood.

NFPA

The National Fire Protection Association (NFPA) publishes standards and codes that represent the consensus of a committee of experts. NFPA standards do not have the force of law, unless written into local law by reference or adoption, nor does NFPA enforce or monitor compliance. However, because NFPA standards reflect the national industry standard, fire organizations that do not follow NFPA standards face a potential for liability in the event of litigation. Some of the NFPA standards that apply specifically to the health and safety of the fire service workforce include the following.

- NFPA 1500 *(Fire Department Occupational Safety and Health Program)* This standard is an umbrella document intended to establish a framework for a comprehensive safety and health program. The standard addresses areas such as risk management, training and education, protective clothing and equipment, incident management and rehabilitation during emergency operations, civil unrest/terrorism, facility safety standards, member assistance and wellness program, and critical incident stress program. NFPA 1500 also requires specifically designated rescue crews at the incident scene. This requirement is based on the fact that firefighters are at greatest risk of injury or death while operating at the scene of an emergency, and that one of the most effective mechanisms for reducing that risk is to

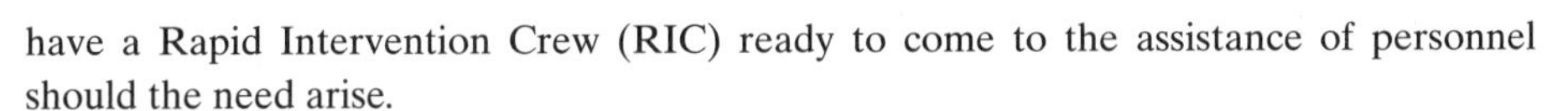

have a Rapid Intervention Crew (RIC) ready to come to the assistance of personnel should the need arise.

The difference between OSHA's two-in, two-out rule and the use of a Rapid Intervention Crew is that, during the initial stages of the incident before sufficient staffing is at the scene, one of the two outside people can be performing other non-critical tasks, as long as they can immediately leave that task to perform a rescue, if necessary. A Rapid Intervention Crew, on the other hand, is dedicated to firefighter rescue, and performs no other task while observing and listening for firefighters potentially in trouble.

- NFPA 1521 *(Fire Department Safety Officer)* This standard differentiates the qualifications, authority, and roles and responsibilities between the Health and Safety Officer and the Incident Safety Officer.
- NFPA 1581 *(Fire Department Infection Control Program)* This standard addresses measures to provide infection control practices and is compatible with all CDC guidelines and regulations. It stresses that a proactive infection control program is necessary for fire departments even if they do not provide EMS, as responses to domestic violence, hazardous materials, and even routine structural fires have the potential for infecting a fire department member.
- NFPA 1582 *(Medical Requirements for Fire Fighters and Information for Fire Department Physicians)* This standard covers the medical requirements necessary for persons who perform firefighting tasks and applies to both candidates and current firefighters. The standard creates two categories of medical conditions. Category A identifies conditions that would prohibit the person from performing firefighting operations. Category B identifies conditions that must be evaluated on a case-by-case basis. The strategic plan should include human resource key indicators and improvement objectives to assess workforce safety and health. Sources of data to assess safety and health indicators include corrective maintenance records, infectious disease exposure reports, back injuries, assaults/exposure to violence, sick time, and workman's compensation. At the time of this writing, this standard is being revised.

◆ WORKFORCE SATISFACTION

Personnel well-being and satisfaction are critical to the organization's delivery of high-quality services. The leadership of the organization must maintain a work environment where well-being factors are included in quality management activities. This requires determining what services, facilities, activities, and opportunities will be available to support personal development, job satisfaction, and well-being. The organization may provide or support services such as career counseling, career development, non-work-related education, and/or special leave for family and community service responsibility.

It is just as important that the quality management program include key indicators and improvement objectives to assess workforce morale and motivation as it is to assess workforce safety and health. Sources of information to analyze morale and motivation indicators include grievance proceedings, staff evaluations of leadership and management, use of career development opportunities, use of sick time, and exit interviews.

CHAPTER SUMMARY

A good quality management program requires the fire organization's leaders to carry the goals identified during the strategic planning process into the realm of human resources. The specific quality goals of the strategic plan must be tied to specific human resource goals. To accomplish this, the human resources that are needed to help ensure the success of the strategic goals must be identified. The human resource plan should focus on not only safety and health factors, but on approaches that foster initiative and strong motivation by empowering and enabling the workforce, such as increasing responsibility and decision making opportunities, education, training, and career development.

Human Resources Checklist

[ ] **Determine current human resources practices.**
- [ ] Gather and analyze the following data.
- [ ] Member satisfaction surveys (baseline)
- [ ] Turnover
- [ ] Absenteeism
- [ ] Grievances
- [ ] Safety records
- [ ] Training records
  - Are instructors trained in teaching methodologies?
  - Does the organization provide training in problem solving, team dynamics, communication, and customer service?
  - Does the organization use alternative teaching methods (TV, CD-ROM, etc.)?
  - Does the organization encourage members to take NFA courses (state and resident)?
- [ ] Recognition program
  - What does the recognition program consist of?
  - Are mechanisms in place to recognize a member by more than simply a letter in his or her file?
  - Is there a formal yearly recognition ceremony?
  - Are families notified of member recognition?
- [ ] Use of teams to determine work practices
  - How often does the organization use member teams?
  - For what types of projects?
  - Are there permanent teams?
  - Are the teams allowed to implement their recommendations, or does management make the final decision regardless of member input?
- [ ] Current job descriptions
  - When were job descriptions last updated?
  - Do current job descriptions reflect actual tasks performed?
  - What mechanisms are in place to ensure that members can meet job performance requirements?
  - Did members have input into the development of the job description and performance requirements?
- [ ] Organization communication
  - What is the typical communication method?
  - Are members satisfied with the communication method?
- [ ] Review city/county personnel policies.

[ ] **Determine workforce health and safety current practices.**
- [ ] Organization complies with all relevant OSHA regulations.
- [ ] Organization complies with NFPA 1500, 1521, 1581, 1582.

[ ] **Identify programs to improve any areas found lacking in the assessment.**
- [ ] Use team approach.

Activities

1. List all of the training or education programs your organization conducts for personnel. Review this list. Based on what you have read in this chapter, what other types of training or education should be added to meet the human resource goals of strategic quality planning?
2. Develop an initial member-satisfaction survey. Describe the expected response of members to this survey based on the current culture of your organization. How would you communicate the purpose of the survey to gain acceptance?
3. Describe your organization's use of workforce teams. Does the current approach foster the development of leadership qualities, empower the workforce, and promote worker satisfaction? If not, develop human resource-related goals for this area.

Bibliography

National Highway and Safety Administration. (1997). *A leadership guide to quality improvement for emergency medical services.* Washington, DC: U.S. Government Printing Office.

United States Fire Administration, National Fire Academy. (1999). *Advanced leadership issues in EMS.* Washington, DC: U.S. Government Printing Office.

Data versus Information

◆ INTRODUCTION

This chapter focuses on the Baldrige category of Information and Analysis. The raw material for information is data. When data is collected and arranged appropriately, it creates new information. Information leads to new knowledge. However, most fire service leaders design system changes based not on what has been proven, but on what they *believe* has been proven. Furthermore, it is often impossible to interface information gathered by one fire organization in a meaningful manner with information gathered by other fire organizations. This handicaps the fire service's ability to quantify—and therefore justify—the reasons for adding or changing services. It does not have to be this way.

The data an organization chooses to collect determines the information produced as the result of analysis. The data collected, the methods used to analyze that data, and the information that results from that analysis has a profound impact on the organization's knowledge, design, and operation. Information is the fire organization's primary means of survival.

◆ SELECTING RELEVANT DATA

Every fire organization collects data. The issue is whether that data is relevant or simply a collection of numbers and facts. To justify services and ensure quality, programs must collect and analyze appropriate data. The data collected should be useful not only to the organization collecting it, but to the fire service on a national level.

Data collection serves as the basis for effective management of all the fire organization's programs on a day-to-day basis by facilitating various administrative requirements. These include equipment/supplies inventory management, expenditure monitoring, maintenance of member records, incident analysis, establishing training schedules, and so forth.

Collection and management of data and its transformation into useful information are fundamental to successful quality monitoring of all of the organization's services and programs. Local, state, and federal agencies may have specific requirements and/or guidelines that must be included for some programs such as the infection control program. Therefore, each organization will have individual forms and checklists to record required data.

Data collection also plays a critical role in evaluation. Baseline data collected before a change to an existing program can be compared to later data collected at established intervals and provide a clear picture of the progress in attaining program objectives. Data analyzed for all three of these purposes should be integrated to provide the information used for planning additions or changes to, or deletions from, existing programs and services.

All data elements selected for each of these purposes must be linked to the program objectives and performance indicators. Determining what data needs to be collected should be part of the planning process as program objectives are developed, reviewed, and revised. This may involve modifying the data elements the organization is currently collecting.

In Chapter Four: Strategic Quality Planning, we identified "providing the safest possible environment for personnel operating on the fire ground" as a program objective and used the following as performance indicators to measure that objective.

- The Incident Command System is followed.
- The two-in, two-out requirement is followed in all Imminently Dangerous to Life and Health (IDLH) situations.
- Rehabilitation procedures follow department SOP.
- All department apparatus meet appropriate standards set by the NFPA.
- Apparatus is maintained through a regular preventive maintenance program described by the manufacturer.
- All personal protective equipment meets appropriate NFPA standards.

Examples of data elements to collect to measure these performance indicators would include the following.

- type of incident
- response times
- adherence to the two-in, two-out requirement and established ICS and rehabilitation SOPs
- maintenance problems of apparatus
- problems with any protective equipment

Data, and the resulting information, must also be reliable, standardized, rapidly accessible, and timely. Reliability is affected by the accuracy of the data collector and the integrity of the storage of the data. Data collection should be automated whenever possible and integrated into work processes. There should be ongoing assessment of data quality. Everyone involved in collecting and entering data should be trained so he or she is knowledgeable about the data being collected and its uses and benefits to the organization.

Data should be standardized. This means that data sets (what is collected), data definitions, codes, classifications, and terminology are uniform across departments within the organization as well as compatible with external databases. A current problem within the fire service is the lack of standardized definitions. This lack of standardization leads to misleading information when benchmarking against other

- Linked to performance indicators
- Reliable—automated whenever possible
- Standardized—use industry-defined definitions for data elements
- Timely—data entered and analyses conducted on a regular basis
- Rapidly accessible—software program allows retrieval of desired information

Figure 6.1 Characteristics of Useful Data.

organizations. A simple example of this is the time the apparatus responds to a dispatch. One organization's definition might be from the time that the apparatus actually receives the call. Another organization might define it as the time personnel clear up on the radio that they are responding. A third organization might define this data element as the time the vehicle actually moves. In order to collect useful information on a national level, organizations should use industry-set data definitions.

In 1976, the National Fire Protection Association (NFPA) developed Standard 901: Uniform Coding for Fire Protection. In 1981, the codes for Fire Service Casualty Reporting, and in 1990, the codes for Hazardous Materials Reporting were developed. The National Fire Incident Reporting System (NFIRS) managed by the United States Fire Administration (USFA) is based on these standards. In 1999, the USFA released NFIRS version 5.0. This version expands the collection of data beyond fire to include the full range of fire organization activity.

Data must also be timely and rapidly accessible. Data collection systems vary from simple, paper-based records to complex multicomputer systems. For data to be timely and rapidly accessible, the organization must computerize the information collected by other means and integrate computer data into a central database. This means the organization must have a person whose primary functions are to keep data up-to-date and accurate, conduct analyses on a regular basis, and submit data to state and regional databases.

Regardless of how relevant the data elements are, performance cannot be measured if the analysis program cannot retrieve the needed information. Many commercial software databases are available, but organizations must work with vendors to customize the software to meet their needs. Software may also be developed at the organizational level, if the computer expertise is present. Figure 6.1 lists the characteristics of useful data.

DATA TYPES AND SOURCES

Incident data are essential to effective quality management. The response form is crucial for evaluating how well the organization is meeting its performance indicators (e.g., response times, adherence to established SOPs, etc.). Incident data also provide valuable information regarding demographics of responses, types of incidents, conditions associated with the incident, and so forth.

Again, using our example of "providing the safest possible environment for personnel operating on the fire ground" and its performance indicators, data sources would include the following.

- incident reports
- ICS forms
- tactical worksheet(s)
- Incident Action Plan
- Site Safety Plan
- SOPs
- post-incident analysis
- apparatus maintenance records
- reports of equipment failure

When submitted to a national database such as the National Fire Incident Reporting System (NFIRS) managed by the United States Fire Administration, incident data also provide the basis for determining national trends. Identifying national trends is critical to determining where the focus and resources of the fire service should be directed.

In 1995, 39 states and the District of Columbia were participating in the NFIRS database. Thirty-nine percent of the more than 33,000 fire departments in those states were providing data (United States Fire Administration, 1995). According to the U.S. Fire Administration, National Fire Data Center, there are currently 12,879 entities reporting NFIRS data, and within each entity, there may be several stations reporting (USFA, personal communication, February 7, 2001).

The National Fire Incident Reporting System (NFIRS) is an important tool used by fire departments to track local fire data and information. It is also designed to be used by state agencies and the U.S. Fire Administration to identify national trends and to initiate fire protection and prevention measures on a large scale. *(Photo by Federal Emergency Management Agency, United States Fire Administration.)*

It is also critical that all data elements be completed. In 2001, the NFIRS Technical Support Center reported that from May through December 2000, they had followed up on over 1100 incomplete incident reports from the 1998 database (B. Morton, personal communication, December 15, 2000). It is difficult at the least and misleading at the most when national trends are analyzed based on adjusted percentages in categories with unknowns of 40% and greater.

The term *stakeholder data* refers to information from customers using (or needing) services from the fire organization. For example, new industry may result in the need to respond to a hazardous materials incident. A new managed care organization may require changes in the current EMS transport protocol. Increased construction or new structure type (e.g., high rise) may require additional numbers or types of apparatus. New construction may also require an increase in available water supply. By collecting and updating pertinent information on community demographics, the organization can predict and improve its ability to respond to stakeholder needs by such things as acquiring new equipment or providing needed training.

Satisfaction data is used to determine how well the organization is meeting the needs of customers and other stakeholders. For example, are responses considered timely? Are inspection personnel (or any personnel, for that matter) efficient, effective, helpful, and courteous? Do personnel work well with other agency personnel involved in incidents (i.e., law enforcement, private EMS provider, etc.)? Is interagency communication timely and effective? It is often not feasible or appropriate to collect this type of data at the time of the incident, but there should be a data collection method in place to ascertain satisfaction data, such as follow-up questionnaires or protocol for follow-up phone calls.

Satisfaction data also applies to internal customers—the members. Data related to sick time, absenteeism, and grievances filed are useful in identifying potential problems with member satisfaction. Although rarely used, an excellent way to determine member satisfaction is to simply ask the member. Because this is not commonplace in fire organizations, a fear of reprisal may exist when the measure is first implemented. This type of program may generate better results if an anonymous questionnaire or an anonymous suggestion box is used.

To be useful, these approaches should have the requirement that any suggestion or concern expressed be accompanied by a potential solution. Not only does this help eliminate use of these methods as a "gripe" mechanism, it provides a valuable source of new and innovative ideas for the organization's leaders. With time and acceptance, the organization can move to the use of member focus groups and interviews.

Process data is important for identifying needs and managing organizational aspects such as finances, training, vehicle and equipment use, age, and replacement, maintenance, and staffing. Process data are also used to determine the root cause of problems. For example, an SOP regarding the location of an Automatic Electronic Defibrillator (AED) on an apparatus is written by a battalion chief who does not ride or function from that apparatus. A review of incident records shows the response time to defibrillation of the patient is not acceptable, even though the time of arrival on scene is. Analysis of process data in conjunction with incident data and input from apparatus personnel may reveal that the designated location of the AED makes retrieval difficult and time consuming. By changing the location based on member input, the response time to defibrillation improves to the acceptable time range. Figure 6.2 summarizes the different types of data.

FIGURE 6.2 ◆ Types of Data.

- Incident Data—identifies demographics of responses, types of incidents, conditions associated with the incidents, etc.
- Stakeholder data—identifies needs of customers
- Satisfaction data—determines how well the department is meeting the needs of customers
- Process data—identifies needs and management aspects such as finances, training, vehicle and equipment use, age, and replacement, maintenance, staffing, etc.

◆ DATA MANAGEMENT

Data management is the key to continuous monitoring and improvement of the quality of the services delivered by the organization. Effective data management begins with the determination of what data should be collected. The data set should focus on information needed to evaluate the effectiveness of the goals and objectives identified in the strategic plan. Every currently collected data element should be reviewed to determine if it is the best source for evaluating and determining the source of problems for each of the performance indicators identified for the objectives. Current data elements may need to be modified or discarded and replaced with new data elements.

If a decrease in staffing resulted in apparatus being placed out of service to meet minimum staffing requirements, additional data elements might be added to allow for the determination of a change in fire loss if the first due company was out of service. Another example is based on the requirement that Rapid Intervention Crews (RIC) are available at all working incidents for immediate assistance to firefighters in trouble. Data elements may be added to determine how often RICs are established, how often they are sent in to assist firefighters in trouble, and the specific reason the RIC is sent in.

The content and frequency of analysis reports should be determined in the planning stages. These may need to be modified at a later date, but the organization needs a starting point. The content and frequency of reports determine how the software should be programmed for analysis and retrieval and how useful the data will be for quality management activities. Irrelevant, incomplete, or out-of-date information is of little value in assessing organizational effectiveness.

Computer software that meets reporting requirements is vital for successful data collection and analysis. Software can be developed internally if the expertise is available, or it can be purchased commercially. If it is purchased commercially, the organization must work closely with the vendor and clearly communicate the needs of the organization so the information sought is actually retrievable in the form desired. The U.S. Fire Administration also provides free software to departments participating in the NFIRS 5.0 national database. Information on the free software is available through the USFA web site at http://www.nfirs.fema.gov/NFIRS_home.htm.

The software program should perform checks for data quality and require that data anomalies and missing entries be resolved and records closed before they are transferred to a central or statewide database. The software should have a "built-in" analysis of unit performance, such as time intervals.

Regardless of where the organization obtains software, the staff using it must be trained and technical support must be available. One person should be responsible for managing the database, including assuring data quality and completeness; running "built-in" analyses; writing ad hoc programs for necessary specific analyses; preparing the analyses used to support organizational planning and evaluations; and submitting data to the state/regional/national database.

The organization should conduct ongoing evaluation of all data management activities. Ongoing evaluation will identify areas needing further development, personnel needing additional training, and equipment necessary to improve productivity.

◆ DATA QUALITY

The usefulness of the information gained through data analyses hinges on the quality of the data submitted to the database. The workforce collects the majority of all data used by fire organizations. Training the workforce has a significant impact on the quality of data. They should understand all of the operational definitions used, the applications, and importance of the data. The workforce should be educated about the process and activities involved in converting the data they collect to useful information. They should also be given an explanation of what that information will be used for, including how the information is linked to the strategic plan.

Procedures for collecting and recording data should be specific, well-defined, and reflect an understanding of the working environment of the members who collect the data. Direct electronic data entry from a laptop or computer in the engine house is the most desirable method, as it permits checks for data completeness shortly after it is collected while details are fresh, increases the likelihood of correcting errors, and reduces data omission.

Other factors that can increase data quality include evaluating the members' knowledge of data collection procedures and operational definitions, early review of submissions by personnel at the central database, continuing education on data quality, and feedback to members. Feedback should be given related to the quality and completeness of reports submitted. This feedback should not be used as a disciplinary tool. Once established as a disciplinary measure, documentation can become very "creative." Inaccurate data results in planning decisions being made on totally erroneous information. Feedback should also be given to members regarding the overall findings of the analyses and compliance for performance indicators.

◆ ORGANIZATIONAL USES OF DATA

Use of data at the organizational level differs from the use of data related to individual performance and even division performance. Organizational data should be related to quality, customers, markets, and operational performance. Combined with financial information, these data are integrated and analyzed to support organization-level review, action, and planning. This section discusses four of the major organizational uses of data (Figure 6.3).

UNDERSTANDING CUSTOMERS AND MARKETS

Fire organizations serve entire communities and populations at risk. Understanding the demographics and sociocultural features of the community is important for planning all organizational activities. Access to demographic databases such as the U.S. Census is helpful. Local chambers of commerce and Better Business Bureaus may be sources for categorizing and identifying emerging industry trends in the community. Working with local industry representatives and organizations may also be helpful in gathering and analyzing incidence data.

IMPROVING CUSTOMER-RELATED DECISIONS

Understanding the needs of customers requires ongoing communication with the various cultural community groups and industry representatives. Regular communication with citizen groups or a citizen council composed of cultural representatives can provide the information needed to improve the organization's community relations.

IMPROVING OPERATIONS-RELATED DECISIONS

Incidence and demand data are critical for operational planning. Demand- and incident-pattern analysis should be used for refining current operations such as apparatus placement. These data should also be used for long-range planning for future operations, such as adding a station, additional equipment and personnel, or new services. Operational performance indicators and cost should be compared to determine if the organization is providing the highest quality services with the money available.

UNDERSTANDING COMPETITIVE PERFORMANCE

Identifying and understanding the competition for any of the services offered by the fire organization is important to ensure that the organization is responsive to the needs of its customers. Answering the following questions can help focus the organization on quality performance and improvement.

- Who is the competition?
- What services do they offer and at what cost?
- What is their level of performance?
- What are the gaps in their services?
- Can and should the organization realign its services to augment their services? To take over their services?

FIGURE 6.3 ◆ Organizational Uses of Data.

- Understand customers and markets
- Improve customer-related decisions
- Improve operations-related decisions
- Understand competitive performance

CHAPTER SUMMARY

Data collection and analysis are central to implementing the strategic quality plan. Getting useful information from data analysis hinges on selecting relevant data elements, training data collectors so data are complete and reliable, generating regular and timely reports, and using the information at the organizational level to improve operational performance and to assure the provision of community-needed services.

Developing and implementing a comprehensive data collection system can be expensive and time consuming. Fire organizations should begin by undertaking those data and information activities allowed within their current resources and develop an action plan to expand their capabilities as necessary.

Data Checklist

[ ] **Select relevant data elements.**

- [ ] Incident Data
  - See Appendix B for a list of National Fire Incident Reporting System (NFIRS) 5.0 data modules. The NFIRS 5.0 Quick Reference Guide is also available online at: http://www.nfirs.fema.gov/refguide.htm. This guide contains the complete description of all elements for each module of incident data.
- [ ] Stakeholder Data
  - Community demographics
  - Changes in community industry
  - New areas of construction and construction type
- [ ] Satisfaction Data
  - Customer follow-up surveys
  - Interagency relationships
  - Use of sick time
  - Absenteeism
  - Grievances
  - Member surveys
- [ ] Process Data
  - Member training
  - Vehicle and equipment use and maintenance
  - Revenue flow
  - Apparatus/equipment replacement
  - Certifications

[ ] **Review currently collected data elements and add/discard based on objectives and performance indicators.**

[ ] **Choose data analysis method/software program.**

- [ ] Is the program available through another city/county agency?
- [ ] If purchased new, arrange to consult with vendor to meet needs of organization.

[ ] **Arrange for data input.**

- [ ] Identify person responsible for managing database.
- [ ] Assure technical support is available.
- [ ] Train members on inputting data correctly.
  - Develop reference guide, including standard definitions.
  - Develop procedures for collecting and recording data.
  - Train all members before beginning input.

[ ] **Monitor accuracy of member-submitted data.**

- [ ] Review submission at central database early and regularly.
  - Provide nondisciplinary feedback to members.
  - Provide additional training as needed.

[ ] **Use the data to benefit the organization.**

- [ ] Understanding customers and improving customer-related decisions
  - Changes in sociocultural features of community
  - Demographic changes
  - Emerging industry trends
  - Establishing on-going communication with community cultural groups and industry representatives
  - Involvement in community programs

[ ] Improve operations-related decisions
- Changes in demand and incident patterns
  - Need to relocate vehicles
  - Need for additional personnel/equipment/facilities
  - Changes in preventive maintenance program
  - Need to add new service/alter existing service
  - Identify changes to budget

[ ] Understand competitive performance
- Duplicated services
- Differences in levels of performance
- Partner or takeover services

Activities

1. Develop a performance indicator for each of the four types of data discussed in this chapter (incident, stakeholder, satisfaction, and process). Identify all the data elements you would need to collect to evaluate compliance.
2. Describe the methods you would use to collect each of the data elements identified in Activity 1. Determine how often you would analyze the data.
3. Describe how you would use the analysis of the performance indicators you developed at the organizational level.

Reference

United States Fire Administration. (1995). *Fire in the United States* (10th ed.). Washington DC: U.S. Government Printing Office.

Bibliography

National Highway and Safety Administration. (1997). *A leadership guide to quality improvement for emergency medical services.* Washington, DC: U.S. Government Printing Office.

United States Fire Administration, National Fire Academy. (1999). *Advanced leadership issues in EMS.* Washington, DC: U.S. Government Printing Office.

Basic Methods of Data Analysis

◆ INTRODUCTION

No one method of data analysis will yield every type of answer sought. This chapter will focus on examples of several methods of data analysis that are useful in determining different types of information.

◆ MULTIVOTING

The use of teams to identify key drivers, problems related to those drivers, and potential solutions was discussed in Chapter 4: Strategic Quality Planning. In the initial stages of identification and during the development of potential solutions, teams often brainstorm ideas. This can result in fairly long lists. Multivoting is best used to prioritize and reduce these lists to a few manageable options. Multivoting helps teams focus on problem solving and identifying priorities and allows each member to participate equally in decision making. Multivoting should not be used when trying to reach consensus on a single issue.

THE MULTIVOTING PROCESS

1. List all items on an easel chart. Numbering each item will facilitate recordkeeping.
2. Allocate to each team member a number of votes equal to approximately one-half the number of items on the list.
3. Direct members to vote for the items they believe are most important until they have used their allotted number of votes. They may use all their votes for one item, or vote for several items.
4. Count the votes for each item. Select the top four to six items.
5. Discuss and rank the remaining items. If the team does not clearly establish the top four to six items, remove those items having the fewest votes, and conduct another vote.

FIGURE 7.1 Multivoting Tally Sheet for QI Project Selection.

Topic	Vote Count	Total
1. Inaccurate run reports	√√√√	4
2. Excessive response time	√√√√√√√√√√	10
3. High number of grievances	√√√√√√√√√√√√	12
4. High incidence of equipment failure	√√√√√√√√√√√√√√√√√√	18
5. Excessive apparatus maintenance down-time	√√√√√√√√√√√√√√√√	16
6. Increase in customer complaints	√√√√√√√√√√√√√√√√√√	18
7. Complaints about staffing	√√√√	4
8. Increase in overtime	√√	2
9. Outdated SOPs	√√√	3
10. Decreased attendance at continuing education	√√	2

A show-of-hands vote is the fastest way to conduct multivoting. Team members should keep track of how many times they have voted, and are on the honor system to vote only the number of times allowed. Ballots are the most confidential method of multivoting and ensure that each member votes only the allowed number of times. However, if there is a large number of items, counting the votes may be time consuming and the team may lose momentum while waiting for the count to be completed. Figure 7.1 shows an example of a multivoting tally sheet for a quality improvement project selection.

HISTOGRAMS

Before making changes to improve performance, it is important to know what the performance is at the present time. One way to describe and evaluate current performance is by developing a *histogram.* A histogram is a type of frequency distribution graph, and is best used to visually display the frequency of data occurrence such as response times, specific vehicle maintenance problems, skills completions, and so forth. It provides a picture of data distribution that shows the spread and shape. Histograms group data into defined intervals and display that data according to frequency of occurrence in each interval.

Data distribution can provide clues about variations in work performance. Distributions can be skewed from the center in either a positive (tail of the distribution to the right) or negative (tail of the distribution to the left). Repeating histograms after a change has been implemented allows comparison with initial findings and helps identify improvement. Spreadsheets such as Excel™ will prepare histograms based on data recorded. Figure 7.2 is an example of a histogram showing the overall average response times for first-responding engines.

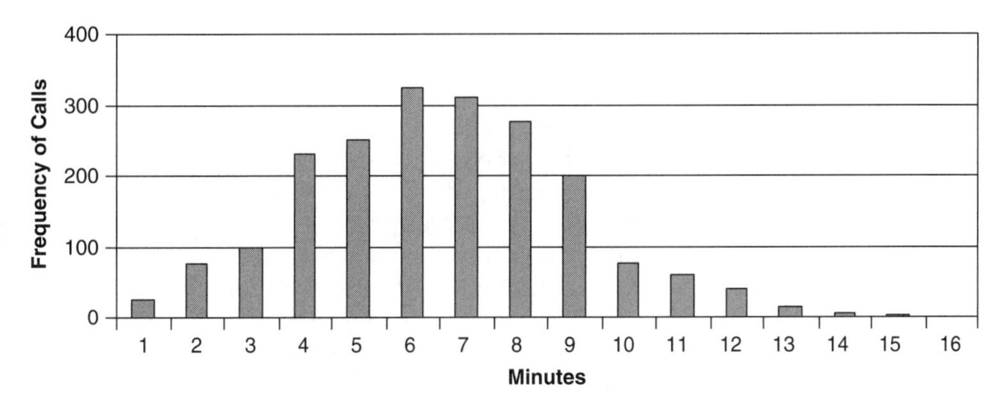

FIGURE 7.2 ◆ First-In Engine Response Times.

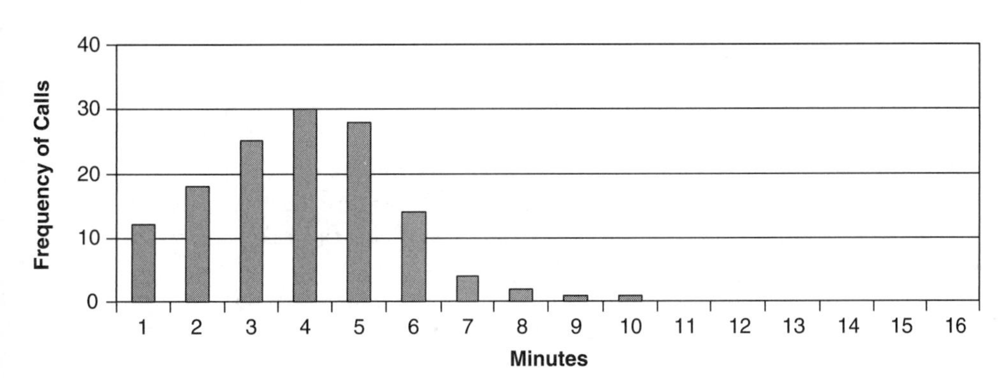

FIGURE 7.3 ◆ Engine 3 Response Times.

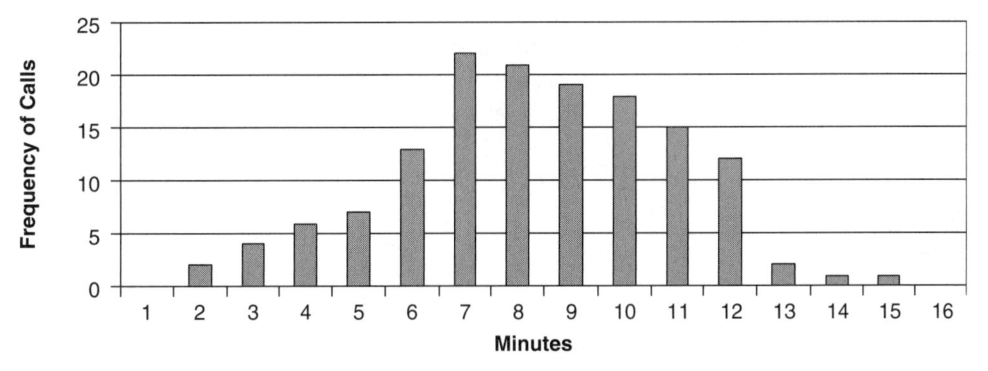

FIGURE 7.4 ◆ Engine 43 Response Times.

This provides valuable data that reveals acceptable overall performance, but extending it to individual engine company response may tell a different story. Figures 7.3 and 7.4 show the response times for two individual engine companies.

The data for Engine 3 demonstrate a positive skew with the majority of response times being less than 6 minutes. The data for Engine 43 demonstrate a negative skew, with the majority of response times being greater than 6 minutes. By reviewing individual apparatus response, data may suggest the need to relocate units or revise territories.

FLOW CHARTS

Much of the work conducted by fire organizations flows between divisions, offices, and other organizations. Picturing the whole process is easier if one can see how specific tasks and activities interrelate to complete the overall activity. Flow charts illustrate the activities performed and the flow of resources and information in a process.

Quality management means constantly looking for ways to improve the effectiveness and efficiency of work processes. Flow charts can help identify activities that reduce effectiveness and efficiency. Through the development of a flow chart, activities may be identified as redundant or unnecessary. Some activities may be identified as being performed in sequence when they might be able to be conducted simultaneously. Flow charts allow the organization to focus on problems at various points within the process rather than attempting to change the entire process.

THE FLOW CHART PROCESS

Flow charts are constructed using various symbols as building blocks.

- An activity is signified by a square or rectangle. The step is written inside.
- Decision points in the process are identified with a diamond. Each path emerging from a decision block is labeled with one of the possible answers to a question that is posed at this point in the process.
- Arrows indicate flow or sequence and direction within the process. The arrow leaves the point of output of one activity and goes to the next activity, where it becomes the input.

Flow charts may be high level or detailed. High-level flow charts illustrate how major groups of related activities interact in a process. High-level charts typically show four to seven subprocesses. Figure 7.5 is an example of a high-level flow chart for response to a fire incident. Figure 7.6 expands the high-level flow chart from initial size-up (assuming command), creating a detailed flow chart.

PERCENT DISTRIBUTION

A great deal of data are analyzed using percents. Performance standard compliance is normally set using a percentage basis. Percent distribution can be used to analyze and determine compliance of specific data components in a performance indicator and can be demonstrated easily in graphic form using spreadsheet applications such as Microsoft Excel™. Let us say you want to determine the most frequently missed elements of incomplete records. You identify the data elements to collect. For this example, we will use the following.

1. Response address
2. Zip code
3. Type of property involved
4. Ignition factor
5. Status of smoke detector

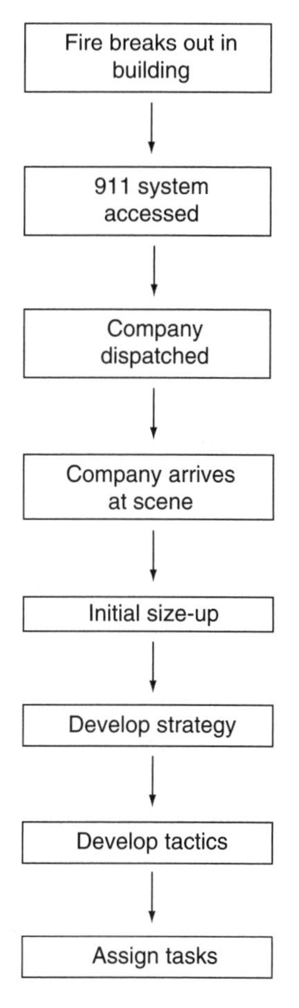

FIGURE 7.5 ◈ High-Level Flow Chart for Fire Incident.

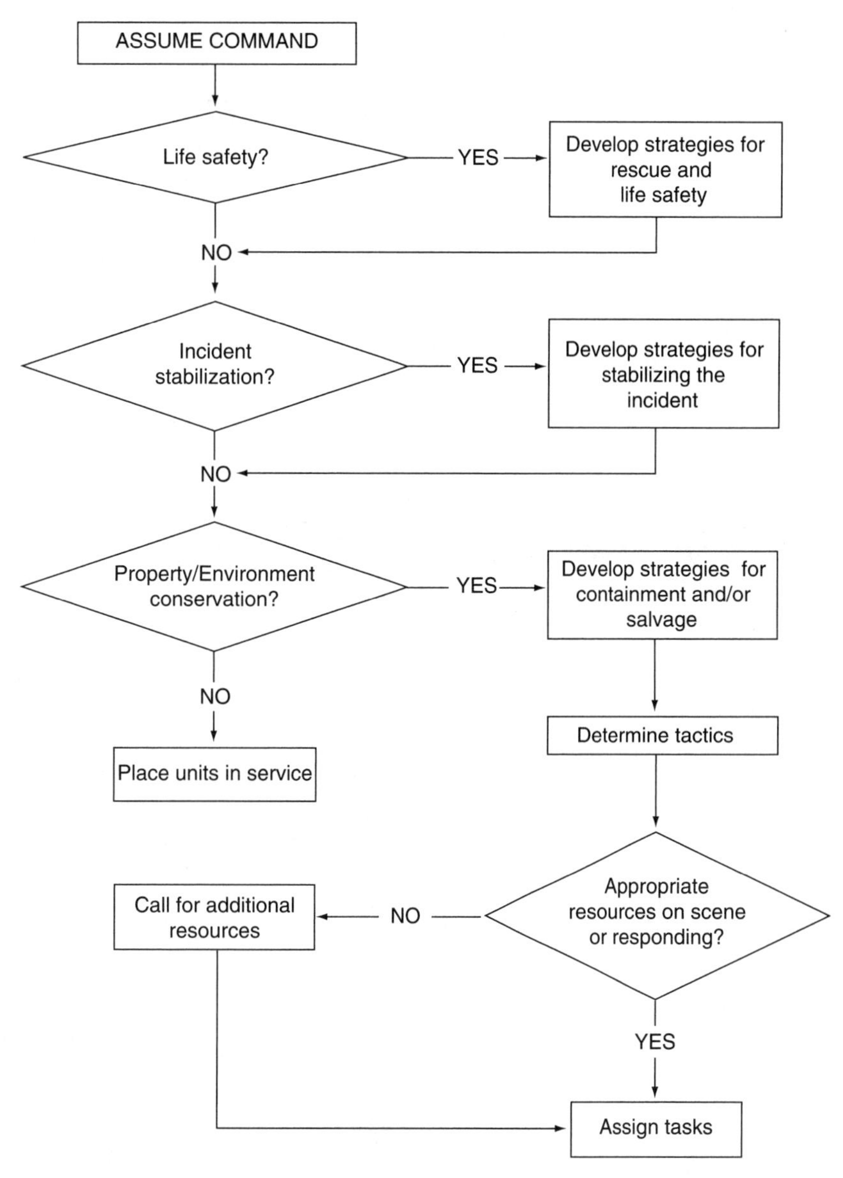

FIGURE 7.6 ◈ Detailed Flow Chart of Fire Incident.

FIGURE 7.7 ◆ Excel Worksheet to Calculate Percentages.*

Address	5	0.04065
Zip code	39	0.317073
Property	23	0.186992
Ignition	10	0.081301
Smoke Detector	62	0.504065

* Formula for calculation: = (B_x/123)

FIGURE 7.8 ◆ Run Report Missing Elements.

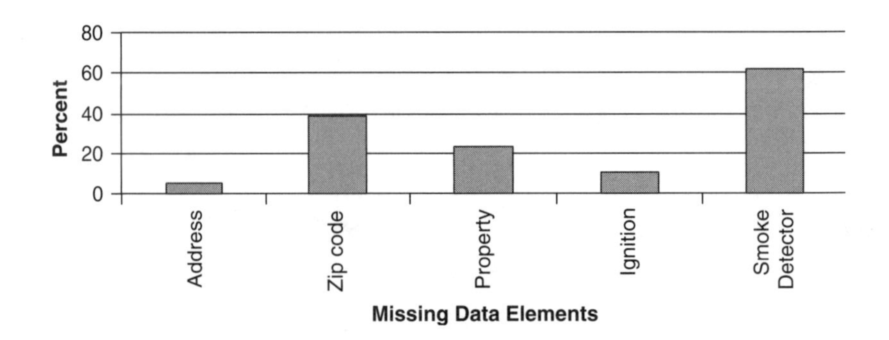

You review a total of 123 responses with missing data elements. Figure 7.7 shows this data entered into an Excel™ worksheet; Figure 7.8 graphically demonstrates that the most common data elements missing are smoke detector status and zip code.

◆ TREND ANALYSIS

Tracking and analyzing data over time can provide valuable information regarding trends. For example, you want to know if workforce satisfaction has improved since implementing self-directed teams and a new recognition program. Indicators you may choose to evaluate workforce satisfaction include both soft and hard data. Soft data might include a member survey with items related to fostering career development/education and facilitating open communications. Hard data include, but are not limited to, sick leave, grievances, and attrition, based on specific items on exit interviews.

People leave jobs for a variety of reasons other than dissatisfaction with the work environment, including medical conditions unrelated to work, the relocation of a spouse or significant other, higher pay, and promotion. Therefore, all reasons for leaving the job may not be applicable for determining worker satisfaction. For our examples, we will use the hard data of grievances filed in a year and the number of total sick days used per year. We will also use soft data gathered through a member survey on the indicators of fostering career development/education and facilitating open communication.

The examples analyze four years' worth of data. Ideally, the organization would have baseline data for the year preceding the implementation of the new program to include for comparison. Graphic representation of analysis may show an obvious trend in a given direction. When there is a less obvious trend, spreadsheet programs

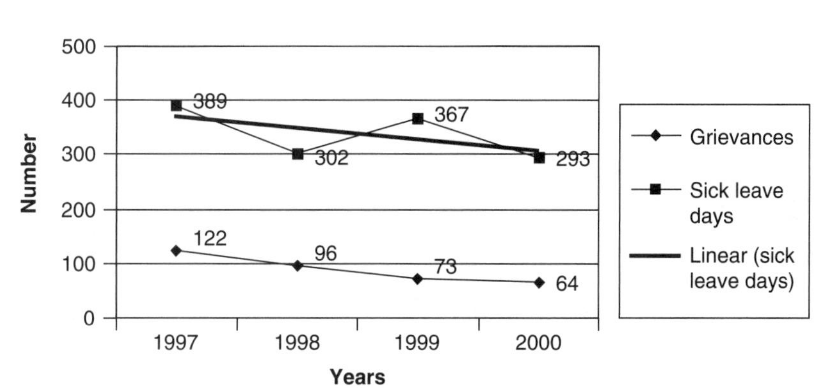

FIGURE 7.9 ◆ Member Satisfaction.

have the capability of inserting a trend line based on the data entered. Figure 7.9 shows an example of analysis of the hard data of grievances filed and sick leave days used for the four years. The decreasing trend in grievances filed is obvious; however, the trend for sick leave use is not quite so obvious, so a trend line has been added to the graph.

Let us say you used a yearly member survey that includes the indicators identified above. The survey gives the respondents five potential answers in regard to the organization's performance on the indicators: poor, fair, average, good, and excellent. Before you can determine a trend, you must analyze each year's data based on the five answer choices. The simplest method for analyzing these yearly data is percent distribution. Table 7.1 shows the distribution for the years we will use in our example.

These yearly results can then be analyzed to establish a trend. Figures 7.10 and 7.11 are the graphic percentage distribution of these yearly results.

There are some cautions that must be remembered when interpreting the results of data analysis. First, if data are collected through a survey, the percentage of returned surveys is critical to the validity of any conclusion. If only 15% of the workforce returns the survey, the validity of results should be questioned. It may be that the 15% who returned the survey are the most dissatisfied with conditions, or it may be that the 15%

TABLE 7.1 ◆ Member Satisfaction Survey Percentage Results

	Poor	Fair	Avg	Good	Exc
Career Development					
1997 (n = 130)	13	26	51	7	3
1998 (n = 189)	10	26	49	12	3
1999 (n = 269)	6	18	52	19	5
2000 (n = 250)	4	14	48	27	7
Communication					
1997 (n = 130)	33	24	39	3	1
1998 (n = 189)	30	29	38	2	1
1999 (n = 269)	26	19	47	5	3
2000 (n = 250)	12	23	42	17	6

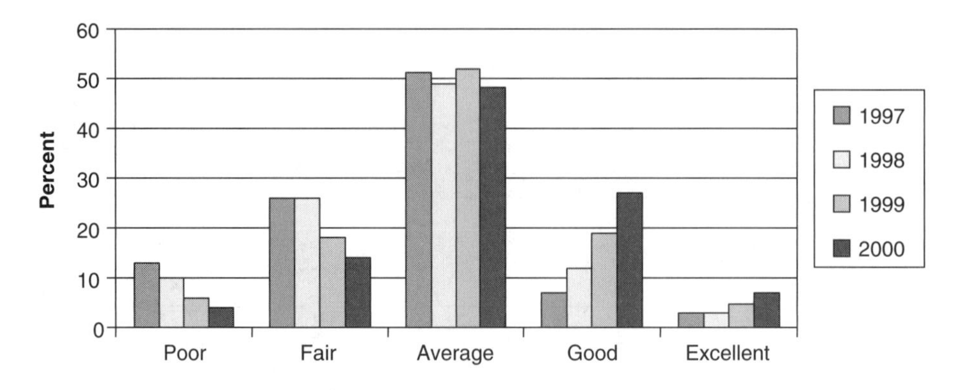

FIGURE 7.10 ◆ Career Development.

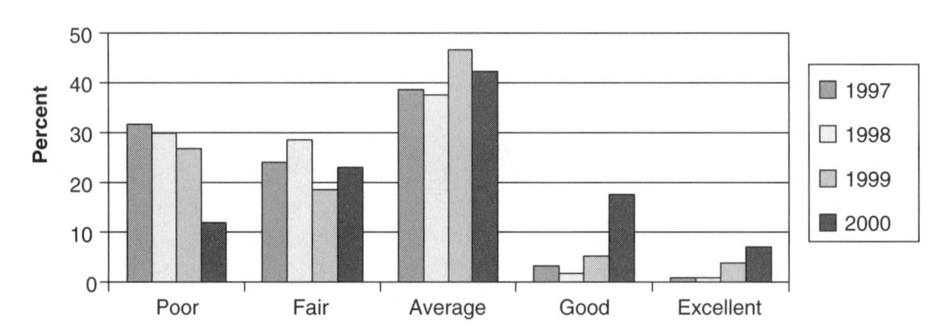

FIGURE 7.11 ◆ Communication.

are the most satisfied. Either way, it is not a large enough sample to provide the basis for evaluation of programs or to make decisions regarding those programs.

Second, the results of data analysis do not yield reasons for the findings. If the response is not what you expected or improvement is not noted, further analysis must be done to determine reason(s) and a new plan should be developed and implemented.

◆ NATIONAL REPORTS

The United States Fire Administration (USFA) and the National Fire Protection Association (NFPA) publish a variety of reports related to fire operations. These reports range from national statistics related to fire incidence to technical reports on a specific fire incident. NFPA has provided a wide range of statistical and data services since the mid-1980s. They provide information in a variety of forms: annual statistical reports, packages of previously published incident descriptions, and customized reports. Some NFPA reports are available free of charge. Complete information about NFPA's reports and services can be found at the organization's website: http://www.nfpa.org/Research.

The United States Fire Administration (USFA) also publishes information in a variety of formats. USFA provides information analysis reports of national trends and specific incidents, as well as a variety of manuals with information for developing and managing various aspects of fire and EMS operations. USFA publications are free and a complete list is available at their web site: http://www.usfa.fema.gov/usfa.pubs. NFPA and USFA analytical reports of fire trends serve as useful tools for fire organizations to benchmark performance standards.

CHAPTER SUMMARY

Fire service organizations can use a variety of methods to analyze data. This chapter has concentrated on very basic methods. If organizations have access to a statistician or a person knowledgeable in statistical methods, data may be analyzed on a higher level using specific statistical tests. These tests are beyond the scope of this book, but readers may choose to access other readings that explain statistical analysis. It is important to choose an appropriate method of analysis if the information is to be useful. Choosing an appropriate method should begin by asking one central question: What do we want to know? Reviewing national reports such as those published by the United States Fire Administration and the National Fire Protection Association can serve as benchmarks for some operational indicators used by the local organization.

Activities

1. Select one month of your organization's engine responses. Prepare a histogram for the overall response times. Then select two engines from different population concentrations and prepare histograms for their response times for that month. Compare individual engine response times with the overall response times for the month. Do the individual engines reflect the same overall response time analysis? If not, what are the differences?
2. Select one work process in your organization. Make your choice a fairly simple one for this exercise. Develop a detailed flow chart for this process.

Bibliography

Gravetter, F., & Wallnau, L. (1996). *Statistics for the behavior sciences* (4th ed.). New York: West Publishing Co.

National Highway and Safety Administration. (1997). *A leadership guide to quality improvement for emergency medical services.* Washington, DC: U.S. Government Printing Office.

Process Management and System Results

◆ INTRODUCTION

Community financial constraints challenge fire organization leaders to design new methods of operating and new services that meet the demands for quality and cost efficiency. This chapter addresses the last three Baldrige categories: Process Management, System Results, and Customer Satisfaction. It reviews how key processes are designed, managed, and improved to achieve higher performance, measuring system results, and assessing customer satisfaction.

◆ DESIGNING AND INTRODUCING NEW SERVICES

The quality-oriented fire organization has a well-defined strategy for designing new services and evaluating and redesigning existing services. Strategies should specify how decisions are made to implement both new primary services and new preventative programs. They should also specify how external changes such as regulatory and standard changes and technology changes will be incorporated into work processes. Finally, they should specify the timing and flow of new service proposals so the operations of both external entities and internal divisions affected by the new service are coordinated in support.

The quality evaluation of the new service should specify what variables are to be measured, who is responsible for carrying out the measurement, and when and where it will occur. There should also be specific procedures developed to ensure that proposals for new services are thoroughly reviewed and pilot-tested where necessary. This helps ensure maximum effectiveness and safety for both external customers and the workforce. The design, evaluation, and pilot-testing process itself should be subject to quality management and improvement.

Table 8.1 ◆ Key Requirements

Customers and Other Stakeholders	Organization Processes
◆ Safety and risk management	◆ Recognition of an emergency
◆ Response timeliness	◆ Bystander intervention
◆ System access	◆ System access
◆ Coordination of response	◆ Dispatch
◆ Availability of trained workers	◆ First responder response
◆ Use of technology	◆ Additional apparatus response
◆ Apparatus capacity and utilization	◆ ICS
◆ Supplier capability	◆ Interaction with outside agencies (e.g., law enforcement, support organizations)
◆ Documentation	

◆ FUTURE TRENDS

Fire organizations of the future may play a direct or supporting role in the delivery of a wide variety of services. These services may involve areas such as health maintenance and promotion (EMS), tactical support for law enforcement in ever-increasing types of violent situations, dealing with new hazardous materials, terrorist events, or providing new safety procedures for the workforce.

Design issues that fire organizations typically must address include the following.

- modification of existing services, such as the introduction of a new technology
- implementation of a new service as the result of research
- new or modified facilities
- new or modified deployment strategies designed to meet operational service requirements
- new or modified work processes to improve productivity or cost efficiency

Design approaches will vary depending on the nature of the service. If the organization is working on several design projects, leaders will need to coordinate resources among the projects. Service design and evaluation must consider the key requirements of both customers and the organization. Table 8.1 summarizes some of these requirements.

◆ SERVICE DELIVERY

Fire organizations are expected to manage delivery of service in a manner that ensures that design requirements are met and that quality, effectiveness, and efficiency are continually improved. Performance indicators help organizations meet that expectation. Regular analysis of performance indicators will reveal patterns of performance. These patterns serve as the basis for quality improvement projects. Once defined, the organization's leaders must select and support the projects based upon criteria relevant to the project's importance within the overall strategic quality objectives of the organization. Indicators that measure specific actions must be developed for each key service.

It is important to understand the difference between quality control and quality improvement in the overall quality management approach. Quality control is taking action to rapidly restore a process to its intended quality level. Quality improvement involves action over a longer period of time that results in achievement of new levels of performance. Quality improvement is a summary approach to process management that involves a review of data over time to generate an aggregate measure of performance.

Frequently, an individual firefighter or company officer will observe or experience an event that will trigger a decision to make a correction. For example, an apparatus develops mechanical problems enroute to a destination. This results in an unacceptable response time. The captain immediately calls dispatch to send another apparatus. This is effective quality control of an isolated event.

Continuing with the response time example, the organization would monitor the response times for all apparatus as a group and look for causes of unacceptable variations. With regular monitoring and definition of acceptable variation limits, managers and leaders will know when to act. Solutions are then designed based on the aggregate analysis of how rates differ when a variety of causal variables are controlled. Let us assume there is a sudden, unacceptable increase in overall response times for apparatus. Potential causes might include time of day, a sudden change in traffic patterns, road construction projects, increased mechanical problems with apparatus, or an annexation that results in extending response territory size.

As an immediate measure, the organization may decide to temporarily relocate units or revise running territories to restore acceptable response times. If this occurs, quality control has been achieved. If the organization designs a project that results in increasing the facilities, apparatus, and personnel as the intervention and that results in decreasing overall response times below the previously acceptable level that is sustained, then quality improvement also has occurred.

Quality management involves every aspect of the organization's operations, including those that support delivery of services. Table 8.2 summarizes some of the support services of fire organizations. Through careful attention to the needs of those who use support services, these functions can be designed and managed to meet ongoing quality standards and drive improvements.

Community services are population-based services that support the community. Such services might include CPR classes, fire and injury prevention education, permanent combined police/fire community information centers, or supplying standby services such as bike medics at public events. The same strategies (objectives and performance indicators, data collection and analysis) that are used in managing and

TABLE 8.2 ◆ Support Services

◆ Recruitment	◆ Accounting/payroll
◆ Human resources	◆ Fleet maintenance
◆ Training	◆ Purchasing
◆ Communications	◆ Materials management
◆ Information systems	

The same strategies that are used in managing and improving the quality and efficiency of firefighting operations can also be applied to public fire and injury prevention programs. *(Photo by Lisa Sachs.)*

improving the quality and efficiency of direct services can also be applied to community services programs.

◆ SUPPLIER PERFORMANCE

Key suppliers are those outside providers that supply the goods and services that are most important to effective functioning of the fire organization. Key suppliers are those who supply key materials (e.g., vehicles, equipment) or services (e.g., occupational health groups, commercial laundries).

The organization typically requires specific commitments in regard to specifications (defined quality levels), delivery times, and price from its suppliers. To ensure supplier accountability, the organization must keep suppliers informed of the organization's ongoing and changing needs and provide the suppliers with feedback as to whether those needs are being met.

Suppliers cannot change their service if the organization does not communicate the changes needed. The use of joint planning, customer-supplier teams, partnerships, streamlined information and data exchanges, long-term agreements, incentives, recognition strategies, benchmarking, and comparative information strengthens working relationships and communication with suppliers.

There are several methods that can be used to implement these approaches. For example, many vendors actively seek departments to field test equipment. This allows the department to "use" equipment free of charge and provides a mechanism for constructive input to the vendor to improve the product to meet organizational needs. Departments should regularly compare notes about vendors' products and performance with other fire departments. These comparisons should include currently used products as well as new products the department is contemplating. Those responsible for procurement should communicate with vendors on a regular basis to provide both positive feedback and concerns. If these measures do not result in acceptable performance, the department may need to change suppliers.

◆ A ROLE FOR THE STATE FIRE MARSHAL'S OFFICE

The State Fire Marshal's Office has the ability to interact with all fire organizations within the state. This office can take a lead role in providing opportunities for interregional comparisons that determine current performance levels and future potential for improvement. The ability to conduct comparisons is very important. However, an individual fire organization can have considerable difficulty gaining access to comparative data from its peer organizations.

The State Fire Marshal's Office can help create a network of leaders who can share strategies for success. This allows the most successful organization in a particular process to work closely with those organizations that are struggling with less success in the same process.

◆ SYSTEM RESULTS

The overall purpose of measuring system results is to determine how well the organization is doing in its key driver areas and the impact of efforts to improve performance in each key driver. Results can be categorized in three major areas (Figure 8.1).

Input results focus on the necessary resource components of the organization. Effective performance by an organization is contingent on the performance of personnel, equipment, administration, and finances. Performance indicators that address the areas related to input include the following.

- Human resource indicators (e.g., safety, workforce well-being, absenteeism, job satisfaction)
- Productivity indicators (e.g., effective use of personnel, materials, energy, information, capital, and other assets)
- Financial performance indicators (e.g., cost of information systems, asset utilization, cost comparisons with similar size organizations)
- Supplier performance indicators (e.g., dependability, availability, durability, and effectiveness of products)

- **Input results:** Focus on necessary resources
- **Process results:** Focus on activities necessary to deliver the service
- **Outcome results:** Focus on the end-result of process activities

FIGURE 8.1 ◆ System Results.

Operation of the fire organization depends on high-quality, low-cost suppliers. Suppliers often have a different organizational structure, lines of authority, and competing priorities than the fire organization. In industries other than the fire service, companies are increasingly requiring data from suppliers on their quality improvement activities as a precondition to contracting for services. In addition to assessing the outcome of their own quality improvement activities, the organization should also be knowledgeable about the quality improvement activities of its suppliers.

Many new technologies, equipment, and procedures are introduced into the fire service based solely on manufacturer-conducted evaluation. The fire service in general needs to be more involved in evaluation and implementation of new technologies to help ensure that these technologies meet the needs of their systems and customers. Actively seeking the status of a field-test site is one method of becoming involved in this process.

Process results focus on the activities involved in delivering the service. Assessment of time intervals in the continuum of delivering service is an important process results measure. Time interval indicators include, but are not limited to, the time the call is dispatched to time of apparatus response (activation time), the time the call is dispatched to arrival at the scene (overall scene response time), and the time from arrival at the scene until a fire is declared under control.

Process results also include the delivery of fire suppression procedures, and hazardous materials response procedures. Results that show trends in success rates, such as a well-functioning ICS, decrease in on-scene injuries, and effective rehabilitation are indicative of how well the strategic quality planning and improvement process is working.

The use of Class A foam and Compressed Air Foam Systems (CAFS) is a relatively new technology for structural firefighting. Many fire departments are investigating the use of these fire suppression agents by examining input results, process results, and outcome results. *(Photo by Bryan Day, Bureau of Land Management.)*

Outcome results are the most important in terms of the fire organization's effectiveness. The most obvious outcome in the area of fire suppression is overall fire incidence. However, other outcomes might include the following.

- Was there a decrease in multiple alarm fires?
- Was there a decrease in lives lost or injuries? This includes both citizens and firefighters.
- Was there a decrease in the dollar amount associated with property loss from fires?
- Is the community satisfied with the results of fire suppression activities? Were those citizens who suffered direct fire loss satisfied with the performance of the organization and their interactions with members?
- Were services provided to the community at the lowest cost possible?
- If the organization begins using Class A foam, is there a difference in the time it takes to control fires?
- If a residence sprinkler ordinance is enacted, what is the change in working house fires in homes with automatic sprinklers?
- If the organization switches to diesel apparatus, what is the change in fuel costs?
- Have technical rescue services resulted in a decrease in extrication times?

Outcomes for EMS focus on the resulting health status of the patient. Although past outcomes have focused on cardiac arrest and trauma, there is overwhelming evidence that EMS has little to no effect on outcomes in these areas, with the exception of early defibrillation. Therefore, organizations providing EMS should also include the following outcomes in assessing effectiveness of EMS.

- Death: Did the patient survive to hospital discharge?
- Disability: Did the patient's functional status improve as a result of the care given?
- Discomfort: Did the patient's symptoms improve?
- Satisfaction: Was the patient/family satisfied with the service rendered?
- Cost: Was treatment provided at the lowest possible cost?

◆ CUSTOMER SATISFACTION

Satisfaction of customers should be a primary concern for fire organizations, yet tracking levels of customer satisfaction is typically one of the weakest parts of fire operations. Information about customer satisfaction levels related to singular events, trend analysis, and comparisons is central to quality management and an important ingredient in strategic quality planning. Unresolved customer satisfaction problems can threaten the stability of the organization in terms of financial support.

Managing the relationship between the organization and its customers requires communication. Customers need easy access to appropriate information and assistance, as well as a way to provide praise or complaints about performance. Organizations can seek feedback from customers in a variety of innovative ways (Figure 8.2).

It is critical that both formal and informal complaints be resolved quickly and effectively. The organization should have procedures for receiving, reviewing, and responding to comments in all their forms. This includes phone calls, e-mails, letters, comments made to workers in the field, calls to individual divisions in the organization, calls or e-mails to mayoral or city "hot lines," and the media. Comments from

Figure 8.2 ◆ Methods for Seeking Customer Feedback.

- Newsletters
- Internet home page postings with e-mail access
- Civic groups
- Focus groups
- Phone numbers posted on the outside of apparatus
- Follow-up survey cards

customers should be managed in ways that build relationships and increase knowledge about specific needs and expectations.

Some customer complaints may be resolved quickly. For example, a citizen calls and complains that an apparatus left ruts in his yard following a response. This can be quickly remedied by sending a city or station crew to repair the damage.

Other problems may surface with analysis of patterns of complaints over time. For example, a sociodemographic or timing pattern of negative feedback may provide clues to potential causes of dissatisfaction. Comparing standards with other similar organizations, or gathering and reviewing data from independent sources such as community groups, may also be helpful in identifying causes. Although fire organizations are in no hurry to expose complaints, emphasizing the perspective that problems and errors are opportunities for improvement increases the likelihood that more organizations will be willing to share comparative data.

CHAPTER SUMMARY

Design approaches to new services vary depending on the nature of the service. Design strategies should include how decisions regarding implementation are made, how external changes will be incorporated, and the timing and flow.

Managing service delivery should focus on continually improving effectiveness and efficiency. Quality management involves every aspect of the organization's operations, including support services. Regular analysis of performance indicators serves as the basis for quality improvement projects. Effective quality management is evaluated by measuring system results.

System results are categorized in three major areas. Input results focus on the organization's resource component. Process results focus on the activities involved in delivering service. Outcome results focus on the end result of service delivery.

Customer satisfaction should be a primary concern. Customer satisfaction can have a significant impact on financial support for the organization. Fire organizations need to handle all customer complaints quickly and effectively and seek innovative ways to solicit customer feedback.

Process Management Checklist

[ ] **Designing/Introducing New Services**

- [ ] Identify how decisions are made (e.g., demographic changes, community feedback, operational analysis, etc.).
- [ ] Identify how regulations and standards will affect the program/service design.
- [ ] Identify how these effects will be incorporated into the service.
- [ ] Identify how changes in regulations, standards, and technology will be incorporated into the service.
- [ ] Identify the impact the new service will have on existing services.
- [ ] Develop a time line that will coordinate the involvement of external entities and internal divisions.
- [ ] Develop the evaluation mechanism for the service.
    - Write the performance indicators.
    - Identify the data elements.
    - Determine the time frame for analysis of data and review of findings.
    - Identify who will develop revisions based on evaluation and how they will be implemented.
- [ ] Conduct a pilot-test, if necessary.
- [ ] Coordinate resources among projects.

[ ] **Service Delivery**

- [ ] Analyze data related to performance indicators regularly.
- [ ] Prioritize projects based on the project's importance and overall strategic quality objectives.
    - Assure performance indicators are developed for each key service of the organization.
- [ ] Regularly assess the effectiveness and need for new community services.
- [ ] Share performance levels and methods for improvements with other organizations.
- [ ] Regularly assess performance indicators related to service delivery.
    - Response times
    - Fire ground procedures
    - Hazardous materials response procedures
    - EMS responses

[ ] **Supplier Performance**

- [ ] Assure specifications are clear and complete.
- [ ] Ask for and review suppliers' quality improvement data before contracting for services.
- [ ] Communicate regularly with suppliers.
- [ ] Become a field-test site.
- [ ] Regularly share information about suppliers with other organizations.
- [ ] Change suppliers if organizational needs are not met after steps listed above are carried out.

[ ] **Outcomes**

- [ ] Assess success and areas where improvement is needed.
    - Decreased fire incidence
    - Decreased multiple alarms
    - Decreased injuries/fatalities (civilian and firefighter)
    - Decreased dollar amount of property lost
    - Services provided at lowest possible cost

- EMS responses
  - Patient survival to hospital
  - Functional status improved as result of care
  - Symptoms improved
  - Patient/family satisfied with service
- Customer satisfaction
  - Follow-up phone call
  - City "hot line"
  - Follow-up survey
  - Manage external customer complaints in timely manner
- Internal customer satisfaction

Activities

1. Based on assessment of community needs, your organization has decided to implement a program to place automatic defibrillators on all first-responding engines. Design a plan to implement this service. Include the following in your plan.
   a. Members of the planning team
   b. Regulations that impact the implementation of the program
   c. Implementation steps, including time lines
   d. Performance indicators used to evaluate the program
   e. Data elements used to assess performance indicators
   f. Timing of evaluations
   g. Estimated cost of the program
2. Identify the methods your organization uses to evaluate customer satisfaction. What other methods suggested in this chapter could your department implement? How would you implement these methods?
3. Describe your organization's method for handling customer complaints. Analyze your current method and identify ways that it can be improved.

Bibliography

National Highway and Safety Administration. (1997). *A leadership guide to quality improvement for emergency medical services.* Washington, DC: U.S. Government Printing Office.

◆◆◆ Appendix A

Self-Assessment of Progress

Material for this self-assessment is drawn from: Brown, M. G., Measuring up against the 1995 Baldrige Criteria. *The Journal for Quality and Participation* 7(7), pp. 66–72. Reprinted with permission of the Association for Quality and Participation, Cincinnati, OH © 1995. All rights reserved. For more information contact, AQP at 513-381-1959 or visit http://www.aqp.org.

The Chief of the Department or a member of the leadership team responsible for developing the organization's focus on quality should complete this self-assessment. When you can answer "yes" to all of the questions in a particular stage, your organization is ready to move into the next stage of development.

The organization should make every effort to move forward stage by stage in all of the seven Baldrige areas simultaneously. As you complete the assessment, you will notice that activities in one category reference activities in another category.

CATEGORY 1: LEADERSHIP

Stage I: Building Potential for Success

- ❑ Does the Chief have enough knowledge regarding quality management theory and benefits for the organization to effectively explain and endorse the topics to others in the organization?
- ❑ Has the Chief established a strategic quality planning group OR has an existing group taken on new focus and responsibility with respect to strategic quality planning?
- ❑ Does the Chief or his designee lead the meetings of the strategic quality planning group?
- ❑ Are all other leaders in the organization knowledgeable about quality management theory? Can they effectively explain and endorse it to others in the organization?
- ❑ Does the organization have written mission, vision, and values statements? Are these posted or distributed in a manner that all members can see them?
- ❑ Did all members of the organization have input into the development of the mission, vision, and values statements?
- ❑ Are the Chief and other leaders of the organization developing a systematic approach for evaluating their own leadership effectiveness and involvement in quality management?
- ❑ Are the criteria used by the leaders to evaluate their own leadership compatible with the organization's vision and values statements?

Stage II: Expanding Knowledge

- ❑ Do the leaders effectively communicate the organization's vision and values to all members?
- ❑ Are most of the leaders directing or participating in educational efforts to increase quality management knowledge and awareness throughout the entire organization?
- ❑ Have the leaders supported the implementation of programs that demonstrate the organization's community citizenship (e.g, fire prevention and safety educational programs, public CPR courses)?

Stage III: Integration and Commitment

- ❑ Has the leadership restructured operations to promote a constant focus on efficiency, high performance, and meeting the needs of both internal and external customers (if necessary)?
- ❑ Do the leaders take an active role in regularly reviewing all performance measures related to strategic quality planning goals and objectives?
- ❑ Is the organization active in general community support activities (e.g., collecting for charitable organizations, collecting toys at the holiday)?

CATEGORY 2: INFORMATION AND ANALYSIS

Stage I: Building Potential for Success

- ❑ Are the data collection and reporting systems designed around the needs of those who use the data to plan and make decisions?
- ❑ Does the data collection strategy identified in the strategic quality plan have a broad focus on information needs (including customer satisfaction, member satisfaction, financial performance, service quality, supplier performance, and operational performance)?
- ❑ Has the organization's ability to collect data and process information for each key performance indicator listed in the strategic quality plan been assessed?
- ❑ Did the assessment result in developing objectives directed at improving the availability and reliability of data used in key performance indicators?

Stage II: Expanding Knowledge

- ❑ Do organization members at all levels understand the correlation between different types of measures of key performance objectives and customer satisfaction and financial performance?
- ❑ Has the organization successfully collected and analyzed data on at least several key performance indicators? Has the information been shared with all members of the organization on a regular basis?
- ❑ Has the organization continued to question all levels of members about how better to meet their decision-making needs with improved data collection and information processing?
- ❑ Has the organization made plans to collect data that will facilitate comparisons of performance with other organizations providing similar services, especially in the areas of service quality, customer satisfaction, supplier performance, member data, and internal operations and support?

Stage III: Integration and Commitment

- ❑ Has the organization evaluated and made many major improvements in measures and data collection and reporting methods over the last few years?
- ❑ Does the organization regularly collect competitive (if appropriate) and benchmark data on: (1) service quality; (2) customer satisfaction; (3) supplier performance; (4) member data; (5) internal operations and support functions; and (6) other appropriate processes and functions?
- ❑ Does the organization systematically evaluate and improve the scope, sources, and uses of its data?
- ❑ Is the data from all areas in the organization summarized into a few key indices, and results analyzed to identify trends and opportunities for improvement?
- ❑ Is there evidence that key organizational decisions and plans are based on analysis of performance data?

CATEGORY 3: STRATEGIC QUALITY PLANNING

Stage I: Developing Potential for Success

- ❑ Has an initial strategic quality plan been completed?
- ❑ Does the strategic quality plan use the organization's mission, vision, and values statements as key references?
- ❑ Does the strategic quality plan reflect the opinions and feedback of members of the organization?
- ❑ Does the strategic quality plan include a list of internal and external customers and their requirements for quality of services?
- ❑ Does the strategic quality plan describe 12-month goals and objectives for expanding knowledge and use of quality management techniques to all members?
- ❑ Has an initial list of key drivers been developed and included in the strategic quality plan?
- ❑ Does the list of key drivers include at least one key performance indicator for each key driver?

Stage II: Expanding Knowledge

- ❑ Has the strategic quality plan been improved over the initial version?
- ❑ Was the revision to the initial plan based on a thorough analysis of customer needs, competition (if applicable), and potential risks to the organization if internal and external customer needs were not met?
- ❑ Does the revised plan describe the needs of internal and external customers? Is there a clear connection between customer needs and the key drivers?
- ❑ Does the plan identify long- and short-term goals, objectives, and strategies for each performance measure?

Stage III: Integration and Commitment

- ❑ Has the organization evaluated and improved its strategic quality planning process several times over the last several years?
- ❑ Does the plan include specific projections illustrating how performance will compare to benchmark organizations?

CATEGORY 4: HUMAN RESOURCE DEVELOPMENT AND MANAGEMENT

Stage I: Developing Potential for Success

- ❑ Has the organization determined the level of worker satisfaction in multiple areas, including compensation, worker safety, opportunity for self-improvement, and job satisfaction?
- ❑ Has the organization reviewed all its operational goals and strategies to see if adequate human resource support exists to meet the goals?
- ❑ Did the review of human resource needs and worker satisfaction include consideration of the need to improve selection, training, involvement, empowerment, and recognition plans?
- ❑ Does the strategic quality plan have specific quality goals and improvement strategies identified for human resource processes (e.g., hiring, career development, and recognition programs)?
- ❑ Does the organization have a structured training curriculum for all levels and functions of workers? Is that training curriculum based on a thorough analysis of worker training needs?
- ❑ Are training needs derived from an analysis of competencies needed to meet key organizational goals?

- ❑ Does the organization have systematic and effective mechanisms to promote on-the-job reinforcement of skills learned in training?
- ❑ Does the organization tailor the content and method used for training to the audience?

Stage II: Expanding Knowledge

- ❑ Has the organization begun or already implemented a number of innovative approaches to job and work design (such as self-directed teams)?
- ❑ Are there now goals and strategies in place for improving worker satisfaction, safety, health, and ergonomics?
- ❑ Has the organization developed a strategy to evaluate the effectiveness of training programs? Has it begun to evaluate at least some of them?
- ❑ Has the organization identified the needs for special services to workers (e.g., counseling, cross-training, re-training, drug/alcohol treatment, etc.)?

Stage III: Integration and Commitment

- ❑ Does the organization use several different approaches to recognizing and rewarding individuals and groups of workers?
- ❑ Do the workers feel well recognized for their accomplishments?
- ❑ Does the organization evaluate the effectiveness of all the training programs it conducts? Is there evidence of improvements in training programs as the result of this evaluation?
- ❑ Does the organization have a well-defined and multifaceted strategy for providing special services to workers?
- ❑ Does the organization use several methods to measure and improve worker satisfaction? Is there evidence that worker satisfaction has improved as a result?

CATEGORY 5: EMS PROCESS MANAGEMENT

Stage I: Building Potential for Success

- ❑ Has the organization developed a strategy to identify and evaluate all key processes that define or support operations to ensure that critical work functions are designed and operate to meet the needs of internal and external customers?
- ❑ Has the organization completed identifying and documenting via flow charts some of the key processes that define and support operations and that must function properly if internal and external customer needs are to be met?
- ❑ Has the organization begun to identify key indicators based on customer requirements for the key processes documented? Have standards been identified for the measures?

Stage II: Expanding Knowledge

- ❑ Has the organization completed documenting its key processes and identified key indicators and standards based on internal and external customer requirements?
- ❑ Has the organization considered the future needs of internal and external customers and used them as a driver to begin designing new processes to meet those needs?
- ❑ Has the organization thoroughly defined quality requirements for all key equipment, materials, and service suppliers? Have those requirements been adequately communicated to the suppliers?
- ❑ Does the organization require suppliers to have preventive and corrective processes in place to ensure they will be able to consistently meet the equipment, materials, and service requirements of the organization?
- ❑ Are data on key process measures collected on a regular basis? Does the organization use valid control strategies to keep all process measures within standards or acceptable levels?

- ❑ Has the documentation of key organizational processes been expanded to include important support functions within the organization? Is data on process measures collected for which specific standards or goals have been set?

Stage III: Integration and Commitment

- ❑ Does the organization design new and/or improved services and support processes based on a thorough analysis of internal or external customer requirements?
- ❑ Does the design of new and/or improved services and support processes include key indicators that will signal if customer needs are being met?
- ❑ Does the design of new and/or improved services and support processes include implementation strategies, policies, or technology that will control the amount of variation in these processes, as measured by key indicators?
- ❑ Are existing service and support process designs reviewed, tested, and validated by taking into consideration the organization's service performance record, use of services, process capabilities, supplier capabilities, and future requirements of internal or external customers?
- ❑ Does the organization systematically appraise its evaluation process? Does it implement new policies and procedures to improve the evaluation process in an effort to reduce the time between evaluation and introduction of improvements?
- ❑ Does the organization use research, benchmarking, new technology, and information from customers to initiate process improvement efforts?
- ❑ Have the organization's key production and delivery processes been re-engineered or improved in dramatic ways over the last few years?
- ❑ Have any of the organization's support processes been re-engineered or improved in dramatic ways, resulting in improvements in cycle time, productivity, and customer satisfaction?
- ❑ Has the organization implemented cooperative efforts to improve supplier quality, such as contractual incentives or supplier certification programs?

CATEGORY 6: SYSTEM RESULTS

Stage I: Building Potential for Success

- ❑ Are active steps to increase members' focus on achieving quality goals underway?
- ❑ Are demonstration projects that will show all personnel the relationship between quality improvement efforts and quality and service improvement outcomes planned?
- ❑ Do efforts to orient members to achieving quality and operational results emphasize the role of measurement and how these measurements will be used?

Stage II: Expanding Knowledge

- ❑ Do all members understand the purpose and meaning of the organization's increasing focus on quality management and improvement of service quality and efficiency? Are they aware that these results will be clearly measured for the purpose of demonstrating the impact of quality improvement efforts?
- ❑ Have there been some successful demonstrations of the impact of quality improvement efforts on any internal or external service outcomes within the organization?
- ❑ Do plans exist to allow comparison of the organization's quality improvement results with the quality improvement efforts of organizations in other geographic areas or jurisdictions?

Stage III: Integration and Commitment

- ❑ Has the organization shown steady improvement in the quality of services over the last three or more years?
- ❑ Are improvements in results seen on all key indicators used to assess product/service quality?

- ❑ Do the organization's quality results compare favorably to those of other similar fire organizations?
- ❑ Do financial results show significant improvement trends over multiple years and levels of performance?
- ❑ Do trends indicate gains in reducing cycle time in operational or support services?
- ❑ Is there evidence over the last three years that the organization has been able to significantly reduce operational costs without damaging quality?
- ❑ Do measures of member satisfaction or morale show improvement trends and levels of performance?
- ❑ Does the organization have data to demonstrate a three-year or more trend of improvements in quality of service and/or product by all major suppliers?

CATEGORY 7: SATISFACTION OF PATIENTS AND OTHER STAKEHOLDERS

Stage I: Building Potential for Success

- ❑ Has the organization determined how it will continuously evaluate its methods for identifying customer requirements?
- ❑ Has the organization identified a set of improvements in the approaches to building positive relationships with customers? Does the information collected on customers and their specific needs appear useful for decision making on how to increase satisfaction levels?
- ❑ Is the organization developing systems for frequently collecting data on hard measures of customer satisfaction, such as increased public financial support, and soft measures, such as opinion surveys or focus groups?

Stage II: Expanding Knowledge

- ❑ Does the organization segment customers according to common needs and characteristics? Does the organization use multiple methods to frequently determine customer needs and requirements related to products and services?
- ❑ Does the organization have multiple ways to make it easy for customers to seek information, comment, or complain about products or services?
- ❑ Does a formal system exist for tracking and resolving formal and informal complaints in a timely manner?

Stage III: Integration and Commitment

- ❑ Does the organization evaluate and show evidence of continuous improvement in approaches to measuring customer satisfaction over the last few years?
- ❑ Is there data to indicate that all major measures of customer satisfaction show a continually improving trend over at least the last three years?
- ❑ Is there data on all major adverse indicators (e.g., complaints, unpaid bills, legal actions) that show decreasing trends?
- ❑ Is research conducted to project future customers and predict what their key requirements are likely to be? Are customers of competitors also studied over at least the last three years?
- ❑ Does customer satisfaction data for all the organization's products and services show continuous improvement over the last three years?
- ❑ Is the organization's level of customer satisfaction superior to that of other fire organizations?

Appendix B

NFIRS Incident Data Elements

Basic Module	*Structure Fire Module*	*Fire Module*
Identification	Structure type	Identification
Incident location	Building status	Property details
Incident type	Building height	Number of residential living units
Aid given or received	Main floor size	Number of buildings involved
Dates and times	Fire origin	Acres burned
Shift and alarms	Fire spread	On-site materials or products
Special studies	Number of stories damaged by flame	Ignition
Actions taken	Material contributing most to flame spread	Area of fire origin
Resources	Item contributing	Heat source
Estimated dollar losses & values	Type of material	Item first ignited
Casualties	Presence of detectors	Types of material first ignited
Detector	Detector type	Cause of ignition
Hazardous materials release	Detector power supply	Factors contributing to ignition
Mixed-use property	Detector operation	Human factors contributing to ignition
Property use	Detector effectiveness	Equipment involved in ignition
Person/Entity involved	Detector failure reason	Equipment power source
Owner	Presence of automatic extinguishment system	Equipment portability
Remarks	Type of automatic extinguishment system	Mobile property involved
Authorization	Automatic extinguishment system operation	Mobile property type & make
	Number of sprinkler heads operating	
	Automatic extinguishment system failure reason	

Civilian Fire Casualty Module

Identification
Injured person
Casualty number
Age or date of birth
Race
Ethnicity
Affiliation
Date and time of injury
Severity
Cause of injury
Human factors contributing to injury
Factors contributing to injury
Activity when injured
Location at time of incident
General location at time of injury
Story at start of incident
Story where injury occurred
Specific location at time of injury
Primary apparent symptom
Primary area of body injured
Disposition

Fire Service Casualty Module

Identification
Injured person
Casualty number
Age or date of birth
Date and time of injury
Responses
Usual assignment
Physical condition just prior to injury
Severity
Taken to
Activity at time of injury
Primary apparent symptom
Primary area of body injured
Cause of firefighter injury
Factor contributing to injury
Object involved in injury
Where injury occurred
Specific location
Vehicle type
Protective equipment
Protective equipment failure
Protective equipment item
Protective equipment problem
Equipment manufacturer, model and serial number

EMS Module

Identification
Number of patients and patient number
Date/Time arrived at patient
Time of patient transfer
Provider impression/assessment
Age or date of birth
Gender
Race
Ethnicity
Human factors
Other factors
Body site of injury
Injury type
Cause of illness/injury
Procedures used
Safety equipment
Cardiac arrest
Initial level of FD provider
Highest level of FD provider on scene
Patient status
Disposition

Apparatus or Resources Module

Identification
Apparatus or resource

Personnel Module

Identification
Apparatus or resource
Personnel section

Hazmat Module

Identification
Hazmat ID
Container type
Estimated container capacity
Units: Capacity
Estimated amount released
Units: Released
Physical state when released
Released into
Released from
Population density
Area affected
Area evacuated
Estimated number of people evacuated
Estimated number of buildings evacuated
Hazmat actions taken
If fire or explosion is involved, which occurred first
Cause of release
Factors contributing to release
Factors affecting mitigation
Equipment involved in release
Mobile property involved in release
Hazmat disposition
Hazmat civilian casualties

Wildland Fire Module

Identification
Alternate location specification
Area type
Wildland fire cause
Human factors contributing to ignition
Factors contributing to ignition
Fire suppression factors
Heat source
Mobile property type
Equipment involved in ignition
Weather information
Number of buildings ignited
Number of buildings threatened
Total acres burned
Primary crops burned
Property management
NFDRS fuel model at origin
Person responsible for fire
Gender of person involved
Age or date of birth
Activity of person
Right of way
Fire behavior

Arson Module

Identification
Agency referred to
Case status
Availability of material first ignited
Suspected motivation factors
Apparent group involvement
Entry method
Extent of fire involvement on arrival
Incendiary devices
Other investigative information
Property ownership
Initial observations
Laboratory used

Juvenile Firesetter Module

Identification
Subject number
Age or date of birth
Gender
Race
Ethnicity
Family type
Motivation/Risk factors
Disposition of person under 18

◆◆◆ Appendix C

Accreditation Programs Related to the Fire Service

While clever marketing may influence purchasing decisions, consumers often look for products that have the Underwriters' Laboratory (UL) label or the Good Housekeeping Seal of Approval. These endorsements indicate that the product meets an established standard of quality and reliability. They also indicate that an unbiased third party has tested the safety of the product.

In the last few years, we have seen the same philosophy applied to the emergency services arena through a concept called *accreditation.* An accreditation program allows for the application of standards from the National Fire Protection Association and the ASTM, as well as the consensus of professionals in the fire and EMS industry. These standards, or benchmarks, indicate the level of efficiency and effectiveness—quality and reliability—that a fire department must reach to be accredited.

Departments working toward accreditation are forced to take a hard look at the processes that define emergency services. This process may be expensive; however, the dollar value of ensuring compliance with a standard set of practices may keep city personnel, emergency service systems, and local governments from suffering a catastrophic failure in emergency services and from significant legal liabilities.

Elected or appointed government officials often struggle to make an accurate assessment of the impact of emergency services on the community. The city manager or mayor is accountable for the efficiency and effectiveness of those services, and should rest easier knowing that an unbiased third party evaluated the emergency services system and found it in line with industry standards.

Two emergency service standards are equivalent to the consumer Good Housekeeping Seal of Approval. The first is the International Association of Fire Chiefs and International City/County Management Association's Joint Commission on Fire Accreditation International (CFAI). The second is the Commission of Accreditation of Ambulance Services (CAAS), which focuses more on the transportation and medical system a municipality deploys. CFAI and CAAS standards are living documents made by members who respond and evaluate their standards continuously.

Several other accreditation programs apply to the fire service. These address educational programs, dispatch centers, and professional qualifications for various officers, including fire chiefs. Some of these programs, along with their Web addresses, are listed below.

Commission on Fire Accreditation International and the Commission of Chief Fire Officer Designation
4500 Southgate Place, Suite 100
Chantilly, VA 20151
(703) 691-4620
Toll free 1-866-866-2324
Fax (703) 691-4622
http://www.cfainet.org

CFAI is designed to evaluate the delivery of fire and emergency services. CCFOD assists in the professional development of fire and emergency services personnel by providing guidance for career planning through participation in the Professional Designation Program.

Commission on Accreditation of Law Enforcement Agencies
10306 Eaton Place, Suite 320
Fairfax, VA 22030
1-800-368-3757
http://www.calea.org
Designed to evaluate and accredit emergency communication centers.

Commission on Accreditation of Allied Health Education Programs
35 East Wacker Drive, Suite 1970
Chicago, IL 60601-2208
(312) 553-9355
Fax (312) 553-9616
http://www.caahep.org
Designed to evaluate and accredit paramedic training programs.

Commission of Accreditation of Ambulance Services
1926 Waukegan Rd., Suite 1
Glenview, IL 60025-1770
(847) 657-6828
Fax (847) 657-6819
http://www.caas.org
Reviews emergency services with a particular emphasis on ambulance transportation.

International Fire Service Accreditation Congress
1700 W. Tyler
Oklahoma State University
Stillwater, OK 74078-8075
(405) 744-8303
Fax (405) 744-8802
http://www.ifsac.org
Provides a comprehensive review of fire service training programs.

Index

M

N

O

P

Q

R